SpringerWienNewYork

88

Fortschritte der Chemie
organischer Naturstoffe

Progress in the
Chemistry of Organic
Natural Products

Founded by
L. Zechmeister

Edited by
W. Herz, H. Falk,
and G. W. Kirby

Authors:
J. F. Grove, E. Reimann, S. Roy

SpringerWienNewYork

Prof. W. Herz, Department of Chemistry,
The Florida State University, Tallahassee, Florida, U.S.A.

Prof. H. Falk, Institut für Chemie,
Johannes-Kepler-Universität, Linz, Austria

Prof. G. W. Kirby, Chemistry Department,
The University of Glasgow, Glasgow, Scotland

This work is subject to copyright.
All rights are reserved, whether the whole or part of the material is concerned, specifically those of translation, reprinting, re-use of illustrations, broadcasting, reproduction by photocopying machines or similar means, and storage in data banks.

© 2007 Springer-Verlag/Wien
Printed in Austria

SpringerWienNewYork is part of
Springer Science + Business Media
springer.com

Product Liability: The publisher can give no guarantee for the information contained in this book. This also refers to that on drug dosage and application thereof. In each individual case the respective user must check the accuracy of the information given by consulting other pharmaceutical literature. The use of registered names, trademarks, etc. in this publication does not imply, even in the absence of a specific statement, that such names are exempt from the relevant protective laws and regulations and therefore free for general use.

Library of Congress Catalog Card Number AC 39-1015

Typesetting: Thomson Press (India) Ltd., Chennai
Printing and binding: Druckerei Theiss GmbH, A-9431 St. Stefan

Printed on acid-free and chlorine-free bleached paper

SPIN: 10975961

With 12 Figures and 1 coloured Plate

ISSN 0071-7886
ISBN-10 3-211-20688-4 SpringerWienNewYork
ISBN-13 978-3-211-20688-1 SpringerWienNewYork

Contents

List of Contributors... IX

Synthesis Pathways to *Erythrina* Alkaloids and *Erythrina* Type Compounds
E. Reimann 1

1. Introduction.. 2
2. Structural Classification of *Erythrina* Alkaloids 4
3. New *Erythrina* Alkaloids... 18
4. Biosynthesis of *Erythrina* Alkaloids 18
   4.1. Erythrinane Alkaloids... 18
   4.2. Homoerythrinane Alkaloids................................... 20
5. Syntheses of *Erythrina* Alkaloids and *Erythrina* Type Compounds 21
   5.1. Methodical Classification 22
   5.2. Erythrinanes 23
      5.2.1. Final Formation of One Ring............................. 23
         5.2.1.1. Ring C (Route C)................................... 23
            5.2.1.1.1. Cyclization of *N*-Phenethylhydroindole
                      Derivatives (Route C(a)) 23
            5.2.1.1.2. Cyclization of Angulary Arylated Hydroindole
                      Derivatives (Route C(b)) 29
         5.2.1.2. Formation of Ring B (Route B) 32
            5.2.1.2.1. Cyclization of N-substituted
                      C5-Spiroisoquinoline Derivatives
                      (Route B(a)) 32
            5.2.1.2.2. Cyclization of C6-Substituted
                      C5-Spiroisoquinoline Derivatives
                      (Route B(b)) 35
         5.2.1.3. Formation of Ring A (Route A)..................... 35
            5.2.1.3.1. Cycloaddition to Pyrroloisoquinolines
                      (Route A(a)) 35
            5.2.1.3.2. Intramolecular Aldol Condensation
                      of Angularly Substituted Pyrroloisoquinoline
                      (Route A(b)) 37
      5.2.2. Simultaneous Formation of More Than One Ring............. 38
         5.2.2.1. Simultaneous Formation of Rings B
                  and C (Route B/C)................................. 39

5.2.2.1.1. Cyclization of Secondary Diphenethyl-
or (Cycloalkyl)ethyl-phenethylamine Derivatives
(Route B/C(a)) 39
5.2.2.1.2. Cyclization of Tertiary Cyclohexyl-ethyl-
phenethyl-amide Derivatives
(Route B/C(b)) 42
5.2.2.2. Cyclization of N-Substituted 1-Acyldihydroisoquinolinium
Derivatives (Route A/B) 44
5.2.2.3. Cyclization of a Highly Functionalized
Homoveratrylimide (Route A/B/C) 45
5.3. Homoerythrinanes 45
5.3.1. Biomimetic Routes 47
5.3.2. Final C Ring Formation Starting from N-Substituted
Phenylhydroindoles 49
5.3.3. A Ring Formation by [2 + 2] Photocycloaddition
to Pyrrolobenzazepines 50
5.3.4. Simultaneous B Ring Formation/C Ring Expansion
Starting from Spiro-2-tetralones 52

6. Pharmacology 53

7. Concluding Remarks 55

References 56

The Trichothecenes and Their Biosynthesis
J. F. Grove (†) 63

1. Introduction 63

2. The Trichothecenes 64
   2.1. Macrocyclic and Non-Macrocyclic Compounds 64
   2.2. Trichothecene Relatives 90
   2.3. Sources 96
   2.4. Oxygenation Pattern 97

3. Biosynthesis 98
   3.1. Simple Trichothecenes 98
      3.1.1. Mevalonic Acid to Trichodiene 98
      3.1.2. Trichodiene to 12,13-Epoxytrichothecene and Isotrichodermol 101
      3.1.3. Further Oxygenation and Esterification of the Trichothecene
             Nucleus: Biosynthesis of Specific Metabolites 104
         3.1.3.1. Trichothecolone 104
         3.1.3.2. Vomitoxin and Derivatives 104
         3.1.3.3. T-2 Toxin 107
         3.1.3.4. Nivalenol and Derivatives 108
      3.1.4. Trichothecene Biosynthetic Gene Clusters 108
   3.2. Trichoverroids and Macrocyclic Trichothecenes 109
   3.3. Trichothecene Relatives 112

References 113

Melanin, Melanogenesis, and Vitiligo
S. Roy... 131

1. Melanin.. 132
   1.1. Introduction 132

2. Chemistry of Melanin... 134
   2.1. Isolation and Analysis 134
   2.2. Solubilization 135
   2.3. Protein Content 135
   2.4. Carboxylic and Phenolic Function.......................... 136
   2.5. Chemical Degradation 136
      2.5.1. Reductive Methods................................... 136
      2.5.2. Oxidative Methods................................... 137
      2.5.3. Pyrolytic Methods 137
   2.6. Spectroscopic Studies..................................... 138
      2.6.1. UV and IR Spectroscopy.............................. 138
      2.6.2. NMR Spectroscopy.................................... 138
      2.6.3. X-Ray Defraction Study 138
      2.6.4. ESR Study... 139
   2.7. Structure of Melanin 139
      2.7.1. Melanin as Homopolymer.............................. 139
      2.7.2. Melanin as Poikilopolymer 140
      2.7.3. Melanin as Bipolymer................................ 141
      2.7.4. Biophysical Model of Melanin Structure.............. 141
      2.7.5. Structure of Phaeomelanin........................... 143
   2.8. Synthesis of Melanin 143
      2.8.1. Electrochemical Synthesis........................... 143
      2.8.2. Photochemical Synthesis 145

3. Characteristic Biophysicochemical Properties of Melanin 145
   3.1. Interaction of Melanin with Light 146
      3.1.1. Melanin in UV and Visible Light 146
      3.1.2. Melanin in the Photoprotection of Skin.............. 146
      3.1.3. Melanin as Light Screen in Eyes 147
   3.2. Melanin and Its Redox Function 148
   3.3. Binding Complexation and Medicinal Aspects of Melanin 149
   3.4. Use of Melanin for Defence 150

4. Melanogenesis 150
   4.1. Melanogenesis *in vivo* 150
      4.1.1. Melanocytes... 151
      4.1.2. The Characteristics of the Enzyme 152
      4.1.3. Regulation of Melanogenesis 153
         4.1.3.1. Physiological Factors.......................... 154
         4.1.3.2. Organic Sulfur Compounds 154
         4.1.3.3. Metal Ions and Other Chemicals................. 154
         4.1.3.4. Vitamins 154
         4.1.3.5. Hormones 155
         4.1.3.6. Neural Influence............................... 157
         4.1.3.7. Malpighian Cells 157
         4.1.3.8. UV Light....................................... 157

4.2. Melanogenesis *in vitro*	157
4.2.1. Enzymatic Melanin Synthesis	157
4.2.1.1. Rearrangement of Dopachrome	158
4.2.1.2. Polymerization of DHI	159
4.2.2. Non-Enzymatic Melanin Synthesis: Model Reaction	161
4.2.2.1. Udenfriend System: A Model for Mixed Function Oxidase	161
4.2.2.2. Melanin Formation Under Udenfriend Conditions	162
5. Vitiligo	164
5.1. Introduction	164
5.2. Melanocytotoxicity: Antimelanocyte-Antibodies Formation	164
5.2.1. The Immune Hypothesis	165
5.2.2. The Neural Hypothesis	165
5.2.3. The Self-Destruction Hypothesis	165
5.2.4. The Composite Hypothesis	165
5.3. Chemotherapy of Vitiligo	166
5.3.1. Psoralens	166
5.3.2. Psoralen Action and UV Light	166
5.3.3. Psoralen Action on Melanogenesis	167
5.4. Abnormal Biochemical Parameters in Vitiligo	168
5.5. Status of Tryptophan in the Melanogenic System	169
5.6. A Composite Hypothesis on Vitiligo	171
References	171
Author Index	187
Subject Index	201

List of Contributors

Grove, Dr. J. F., 3 Homestead Court, Welwyn Garden City, Herts AL7 4LY, England (deceased)

Reimann, Prof. Dr. E., Department of Pharmacy, Ludwig-Maximilians-Universität, Butenandtstrasse 5–13, 81377 München, Germany,
e-mail: ebrei@cup.uni-muenchen.de

Roy, Dr. S., Institute of Natural Products, 8 J. N. Roy Lane, Kolkata 700006, India,
e-mail: shyamali_radha@yahoo.co.in

Synthesis Pathways to *Erythrina* Alkaloids and *Erythrina* Type Compounds

Eberhard Reimann

Department of Pharmacy,
Ludwig-Maximilians-Universität München, Germany

Contents

1. Introduction	2
2. Structural Classification of *Erythrina* Alkaloids	4
3. New *Erythrina* Alkaloids	18
4. Biosynthesis of *Erythrina* Alkaloids	18
4.1. Erythrinane Alkaloids	18
4.2. Homoerythrinane Alkaloids	20
5. Syntheses of *Erythrina* Alkaloids and *Erythrina* Type Compounds	21
5.1. Methodical Classification	22
5.2. Erythrinanes	23
5.2.1. Final Formation of One Ring	23
5.2.1.1. Ring C (Route C)	23
5.2.1.1.1. Cyclization of *N*-Phenethylhydroindole Derivatives (Route C(a))	23
5.2.1.1.2. Cyclization of Angulary Arylated Hydroindole Derivatives (Route C(b))	29
5.2.1.2. Formation of Ring B (Route B)	32
5.2.1.2.1. Cyclization of N-substituted C5-Spiroisoquinoline Derivatives (Route B(a))	32
5.2.1.2.2. Cyclization of C6-Substituted C5-Spiroisoquinoline Derivatives (Route B(b))	35
5.2.1.3. Formation of Ring A (Route A)	35
5.2.1.3.1. Cycloaddition to Pyrroloisoquinolines (Route A(a))	35
5.2.1.3.2. Intramolecular Aldol Condensation of Angularly Substituted Pyrroloisoquinoline (Route A(b))	37
5.2.2. Simultaneous Formation of More Than One Ring	38
5.2.2.1. Simultaneous Formation of Rings B and C (Route B/C)	39

            5.2.2.1.1. Cyclization of Secondary Diphenethyl- or (Cycloalkyl)ethyl-phenethylamine Derivatives (Route B/C(a)) 39
            5.2.2.1.2. Cyclization of Tertiary Cyclohexyl-ethyl-phenethyl-amide Derivatives (Route B/C(b)) 42
       5.2.2.2. Cyclization of N-Substituted 1-Acyldihydroisoquinolinium Derivatives (Route A/B) 44
       5.2.2.3. Cyclization of a Highly Functionalized Homoveratrylimide (Route A/B/C) 45
  5.3. Homoerythrinanes 45
    5.3.1. Biomimetic Routes 47
    5.3.2. Final C Ring Formation Starting from N-Substituted Phenylhydroindoles 49
    5.3.3. A Ring Formation by [2 + 2] Photocycloaddition to Pyrrolobenzazepines 50
    5.3.4. Simultaneous B Ring Formation/C Ring Expansion Starting from Spiro-2-tetralones 52

6. Pharmacology 53

7. Concluding Remarks 55

References 56

1. Introduction

The history of erythrina research begins at the end of the 19th century. During the last two decades of that time extracts from species of *Erythrina* have been found to exhibit curare-like neuromuscular blocking activities which are caused by alkaloids occurring therein (*1–4*).

It was Altamirano (*2*), who obtained a silky shining, crystallinic acetate as well as Greshoff (*3*) who already isolated several basic unspecified compounds. Because of their remarkable biological activity he suggested a systematic phytochemical examination of the genus *Erythrina*. But it still has taken at least half a century before this has been realized for the first time by Folkers. He has shown that more than fifty *Erythrina* species – as an example, *E. crista-galli* is shown in Plate 1 – are containing the typical alkaloids exhibiting the same curare-like activity reported earlier (*5, 6*). Moreover, his group succeeded in isolating the first crystallized erythrinane alkaloid named erythroidine (**14**; Fig. 2) (*7*). Soon after numerous alkaloids have been isolated, *e.g.* erythramine (**8**) (*8*), erythraline (**3**) (*9*), erythratine (**9**) (*10*), erysodine (**6**), erysopine (**4**), and erysovine (**5**) (*11*). Another decade later the fundamental investigations of Prelog

References, pp. 56–62

Plate 1. Erythrina crista-galli L (Coral Tree)

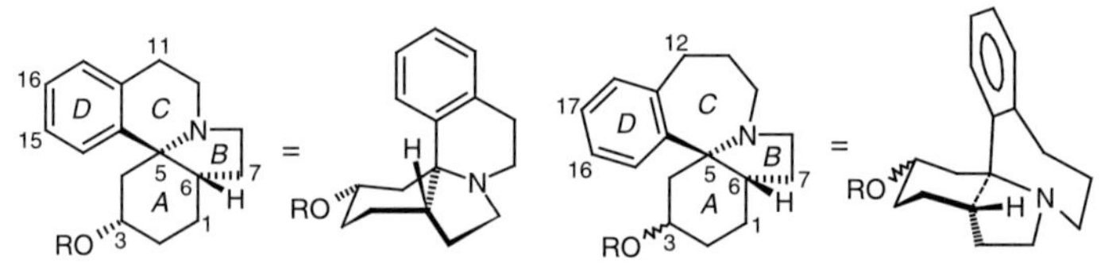

(3R,5S,6S / 3S,5R,6R)-Erythrinane (1)   (5S,6S / 5R,6R)-Schelhammerane (2)
(Homoerythrinane alkaloids)

Fig. 1. Stereostructure and atom labeling of the *Erythrina* alkaloid frameworks

(*12, 13*) and *Boekelheide* (*14*) have finally led to the correct structural framework of the erythrinane alkaloids (parent compound **1**, Fig. 1).

In the late 1960s ring C-homologues of erythrinane alkaloids have been anticipated from the biosynthetic pathway of certain alkaloids, which are known to be generated from 1-phenethyl-isoquinoline derivatives as precursors (*15, 16*). Only a short time later such compounds named homoerythrinanes, homoerythrina alkaloids, or schelhammeranes indeed have been found in the plant kingdom (*17*) (parent compound **2**, Fig. 1).

Due to the increasing attraction and rapid extension in this field the *Erythrina* alkaloids have been regularly reviewed concerning occurrence, structure, analytic and spectral properties, biosynthesis, total synthesis, and biological activities covering the literature up to 1997. The most important reviews are cited in Refs. *18–24*.

The present contribution will give a brief classification of the *Erythrina* alkaloids, a compilation of new alkaloids isolated from 1997 to 2004 covering source, structure, analytical/spectral data, a new pathway of their biosynthesis, an overview of all the synthesis strategies hitherto known for the erythrinane alkaloids including several approaches to the homoerythrinane group, and finally a short review of their biological activities.

2. Structural Classification of *Erythrina* Alkaloids

The erythrina-type alkaloids are characterized by their unique tetracyclic spiroamine framework. They are generally classified into two main groups: Alkaloids predominantly possessing a 6-5-6-6-membered indoloisoquinoline core are called erythrinanes and those exhibiting a 6-5-7-6-membered indolobenzazepine skeleton are generally called schelhammeranes or homoerythrinane alkaloids (see Fig. 1).

Depending on the nature of the D ring both groups in turn may be subdivided into aromatic and non-aromatic alkaloids, the latter of which

References, pp. 56–62

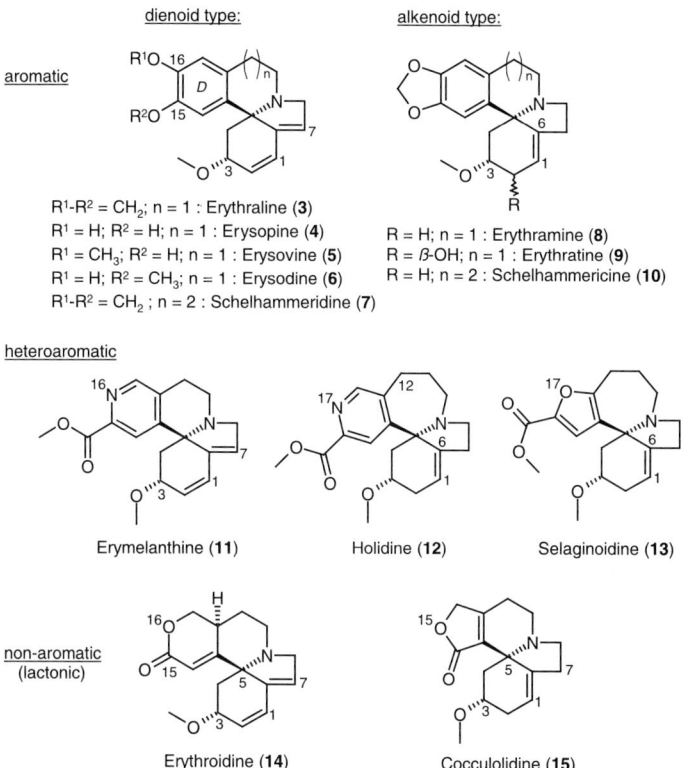

Fig. 2. General classification of *Erythrina* alkaloids: Dienoid and alkenoid type alkaloids and D ring modifications

including ring D oxa-compounds, are usually also called the lactonic alkaloids. In addition, in both series there have been isolated alkaloids containing a pyridyl instead of a phenyl unit, which are known as erymelanthine (**11**) and holidine (**12**) (16-azaerythrinane and 17-aza-homoerythrinane derivatives) belonging to two further different subtypes of these alkaloids (*25, 26*) (see below and Fig. 2). Several D-*seco*-derivatives in the homoerythrinane group should also be mentioned (see *e.g.* **32**, Table 2).

Finally, the typical position and the number of olefinic bonds in the A and B ring have led to a further subdivision into dienoids and alkenoids in both alkaloid series. The former are characterized by a conjugated diene unit covering C atoms 1, 2, 6, and 7, while the latter possess only one double bond in the 1,6-position (see Fig. 2).

The aromatic erythrinanes and homoerythrinanes as the most important members of the *Erythrina* alkaloids show substitution patterns of

Table 1. New Erythrinane Alkaloids

Nr.	Trivial name(s)/ Formula/Structure	M.p./°C ±[α]$_D$/° cm^2 g^{-1}	IR: $\bar{\nu}$/cm^{-1} UV: λ_{max}/nm (log ε/mol^{-1} dm^3 cm^{-1}) MS: m/z	^{1}H NMR: δ/ppm ^{13}C NMR: δ/ppm	Natural source	Ref.
16	Erythrosotidienone C$_{17}$H$_{15}$NO$_3$ (281.31)	250 —	2930, 2860, 1740, 1610, 1460, 1440, 1380, 1280–1260, 1160, 1130, 1075, 1040, 975, 960, 860–840, 730, 720, 675, 645, 610, 580 — 281 (5.0), 280 (22.5), 279 (100), 267 (7.5), 265 (7.5), 253 (17.5), 227 (2.5), 226, 199 (15.4), 174 (15.0), 167 (10.0), 133 (5.0), 81 (31.6), 20 (8.0), 77 (5.0), 76 (80.0), 56 (27.3), 55 (42.5)	6.02 (d, J = 10 Hz, 1-H), 5.94 (s, OCH$_2$O), 5.72, 6.02 (d, J = 10 Hz, 1-H), 5.94 (s, OCH$_2$O), 5.72 (m, 2-H), 3.63 (t, J = 1.5 and 4.5 Hz, 2H, 10-H), 3.27–1.87 (m, 6H) 184.0 (CO), 152.0 (C-16), 152.4 (C-15), 132.0 (C-2), 131.0 (C-1), 129.6 (C-7), 128.3 (C-12), 128.1 (C-13), 114.0 (C-17), 110.1 (C-14), 104.0 (C-6), 102.8 (OCH$_2$O), 68.1 (C-5), 40.3 (C-10), 32.0 (C-3), 30.4 (C-4), 23.2 (C-11)	*E. variegata* Flowers	(33)
17	Erythromotidienone C$_{18}$H$_{19}$NO$_3$ (297.35)	172 —	2920, 2880, 1760, 1600, 1450, 1380, 1280–1260, 1250, 1170, 1110, 1075, 1055, 1025, 975, 960, 925, 885, 835, 800, 765, 730, 700 — 297 (5), 296 (8), 295 (11), 269 (19), 255 (30), 239 (8), 235 (10), 213 (16), 185 (16) 83, 81 (43)	7.32 (dd, J = 2.5/9.0 Hz, 16- or 15-H), 7.21 (d, J = 2.7 Hz, 14-H), 7.13 (d, J = 2.3 Hz, 17-H), 6.62 (s, 7-H), 6.34 (d, J = 9.6 Hz, 1-H), 6.02 (dd, J = 2.5/7.5 Hz, 2-H), 3.86 and 3.36 (2s, 2 OCH$_3$), 3.76–1.92 (m, 7H) ^{13}C not reported	*E. variegata* Flowers	(33)

18	(+)-10,11-Dioxoerysotrine $C_{19}H_{19}NO_5$ (341.36)	174–176	1710, 1680, 1650, 1610	351 (3.54), 292 (sh, 3.62), 247 (4.17), 206 (4.38)	7.48 (s, 17-H), 7.17 (s, 14-H), 6.74 (dd, $J = 11.5/2.4$ Hz, 1-H), 6.00 (d, $J = 11.5$ Hz, 2-H), 5.88 (br s, 7-H), 4.64 (br s, 8-H), 3.95 (16-OCH$_3$), 3.91 (s, 15-OCH$_3$), 3.64 (t, $J = 7.9$ Hz, 3-H), 3.23 (s, 3-OCH$_3$), 2.30 (d, $J = 7.9$ Hz, 4-H)	*E. latissima* (34)
		+167.5		341 (M$^+$, 100), 310 (30), 282 (25), 257 (20)	181.8 (C-11), 159.7 (C-10), 153.5 (C-15), 149.5 (C-16), 141.7 (C-13), 138.0 (C-6), 132.6 (C-2), 124.9 (C-1), 124.2 (C-12), 121.2 (C-7), 111.1 (C-17), 106.4 (C-14), 76.1 (C-3), 70.8 (C-5), 56.9 (3-OCH$_3$), 56.7 (15- and 16-OCH$_3$), 54.7 (C-8), 49.9 (C-4)	
19	11-Acetylerysotrine $C_{21}H_{25}NO_4$ (355.43)	yellow oil	1760 (COCH$_3$), 1600 (C=C)	283.1 (3.7), 230.5 (4.2) 355 (M$^+$)	6.95 (s, 17-H), 6.82 (s, 14-H), 6.65 (d, $J = 10.0$ Hz, 2-H), 6.05 (d, $J = 10.0$ Hz, 1-H), 4.74 (t, $J = 3.4$ Hz, 11-H), 4.05 (m, 3-H$_{ax}$), 3.94 (s, 16-OCH$_3$), 3.85 (s, 15-OCH$_3$), 3.68 (dd, $J = 13.5/3.5$ Hz, 10-H$_{ax}$), 3.32 (s, 3-OCH$_3$), 3.14 (dd, $J = 13.5/6.6$ Hz, 10-H$_{eq}$), 2.42 (dd, $J = 11.5/3.5$ Hz, 4-H$_{eq}$), 2.13 (s, COCH$_3$), 1.87 (t, $J = 11.5$ Hz, 4-H$_{ax}$) ^{13}C not reported	*E. stricta* (35)
		—				

Table 1 (continued)

Nr.	Trivial name(s)/ Formula/Structure	M.p./°C $\pm[\alpha]_D/°$ $cm^2 g^{-1}$	IR: $\bar{\nu}/cm^{-1}$ UV: λ_{max}/nm $(\log \varepsilon/mol^{-1} dm^3 cm^{-1})$ MS: m/z	^{1}H NMR: δ/ppm ^{13}C NMR: δ/ppm	Natural source	Ref.
20	10,11-Dioxoerythraline $C_{18}H_{15}NO_5$ (325.32)	amorph. solid	1710, 1680, 1650, 1610 351 (3.53), 292 (sh, 3.62), 247 (4.17), 204 (4.32)	7.41 (s, 17-H), 7.12 (s, 14-H), 6.68 (dd, $J = 10.3/2.2$ Hz, 1-H), 6.10 (d, $J = 1.5$ Hz, 1H, OCH$_2$O), 6.07 (d, $J = 1.5$ Hz, 1H, OCH$_2$O), 5.97 (d, $J = 10.3$ Hz, 2-H), 5.84 (br s, 7-H), 4.61 (br s, 2H, 8-H), 3.63 (m, 3-H), 3.24 (s, OCH$_3$), 2.25–2.31 (m, 2H, 4-H)	*E. bidwillii*	(36)
		+254	325 (M$^+$, 100), 310 (12), 294 (46), 292 (34), 282 (37), 266 (83), 264 (63), 254 (11), 252 (15), 240 (27), 226 (26), 213 (18), 209 (21), 165 (13), 152 (22)	181.5 (C-11), 159.3 (C-10), 152.0 (C-16 or C-15), 148.0 (C-15 or C-16), 143.5 (C-13), 137.5 (C-6), 132.2 (C-2), 125.8 (C-12), 124.2 (C-1), 120.6 (C-7), 108.6 (C-17), 104.1 (C-14), 102.4 (OCH$_2$O), 75.5 (C-3), 70.5 (C-5), 56.5 (OCH$_3$), 54.2 (C-8), 49.6 (C-4)		
21	8-Oxoerythraline-epoxide $C_{18}H_{17}NO_5$ (327.34)	colourl. oil	1680 289 (3.78), 205 (4.59)	6.72 (s, 17-H), 6.55 (s, 14-H), 6.49 (s, 7-H), 5.98 (d, $J = 1.5$ Hz, 1H, OCH$_2$O), 5.94 (d, $J = 1.5$ Hz, 1H, OCH$_2$O), 4.24 (d, $J = 4.0$ Hz, 1-H), 3.83 (ddd, $J = 12.5/9.5/7.3$ Hz, 10-H$_{ax}$), 3.83 (d, $J = 4.0$ Hz, 2-H), 3.64 (m, 3-H), 3.57 (ddd, $J = 12.5/7.3/3.7$ Hz, 10-H$_{eq}$),	*E. bidwillii*	(36)
		+94	327 (M$^+$, 53), 311 (14), 298 (30), 296 (22), 278 (14), 266 (14), 241 (100), 212 (43)			

22	(+)-Erythbidin B $C_{18}H_{17}NO_5$ (327.34)	amorph. solid +148	3450, 1670, 1650 205 (4.52), 243 (4.04), 291 (3.54) 327 (M$^+$, 100), 312 (23), 296 (74), 294 (34), 284 (6), 278 (13), 270 (7), 268 (12), 266 (14), 250 (5), 240 (7), 238 (7), 227 (6.5), 181 (6.7), 165 (8), 149 (10) CD: $\Delta\varepsilon/\mathrm{mol}^{-1}\,\mathrm{dm}^3\,\mathrm{cm}^{-1}$ (λ/nm) = +3.51 (292), −17.31 (228), +20.76 (202)	3.41 (s, OCH$_3$), 3.09 (ddd, J = 16.1/9.5/7.3 Hz, 11-H$_{ax}$), 2.93 (ddd, J = 16.1/7.3/3.7 Hz, 11-H$_{eq}$), 2.36 (dd, J = 11.7/5.1 Hz, 4-H$_{eq}$), 1.59 (t-like, J = 11.7 Hz, 4-H$_{ax}$) 170.2 (C-8), 155.0 (C-6), 147.2 (C-15 or C-16), 146.0 (C-16 or C-15), 130.0 (C-13), 129.7 (C-7), 128.0 (C-12), 109.8 (C-17), 105.7 (C-14), 101.3 (OCH$_2$O), 74.6 (C-3), 67.4 (C-5), 56.6 (CH$_3$), 52.9 (C-2), 48.7 (C-1), 37.4 (C-10), 32.9 (C-4), 27.5 (C-11) 7.22 (s, 17-H), 6.88 (s, 14-H), 6.63 (dd, J = 10.3/2.2 Hz, 1-H), 5.99 (d, J = 10.3 Hz, 2-H), 5.98 (d, J = 1.5 Hz, 1H, OCH$_2$O), 5.94 (d, J = 1.5 Hz, 1H, OCH$_2$O), 5.76 (br s, 7-H), 5.26 (s, 10-H), 4.43 (d, J = 17.6 Hz, 1H, 8-H), 4.37 (dd, J = 17.6/2.2 Hz, 1H, 8-H), 4.05 (br s, OH), 3.72 (m, 3-H), 3.30 (s, OCH$_3$), 2.60 (dd, J = 11.0/5.1 Hz, 4-H$_{eq}$), 1.95 (t, J = 11.0 Hz, 4-H$_{ax}$)	*E. bidwillii* (37)

Table 1 (continued)

Nr.	Trivial name(s)/ Formula/Structure	M.p./°C ±[α]$_D$/° cm^2 g^{-1}	IR: $\tilde{\nu}$/cm^{-1} UV: λ_{max}/nm (log ε/mol^{-1} dm^3 cm^{-1}) MS: m/z	^{1}H NMR: δ/ppm ^{13}C NMR: δ/ppm	Natural source	Ref.
			CD: $\Delta\varepsilon$/mol^{-1} dm^3 cm^{-1} (λ/nm) = +226 (286), +4.40 (252), −9.16 (218) (MeOH, c 3.06 × 10^{-5})	172.8 (C-11), 147.4 (C-15), 146.7 (C-16), 138.6 (C-6), 131.6 (C-2), 131.2 (C-13), 129.6 (C-12), 124.1 (C-1), 119.9 (C-7), 106.1 (C-17), 103.9 (C-14), 101.4 (OCH$_2$O), 76.1 (C-3), 71.7 (C-5), 67.7 (C-10), 56.4 (OCH$_3$), 54.1 (C-8), 39.5 (C-4)		
23	(+)-Epierythrinine C$_{18}$H$_{19}$NO$_4$ (313.35)	amorph.[a] —	spectra not reported	^{1}H: differentiated signals for the *epi*-isomer: 7.04 (s, 17-H), 6.84 (s, 14-H), 6.54 (dd, 1-H), 6.03 (dm, 2-H), 5.94 (d, J = 1.4 Hz, OCH$_2$O), 5.91 (d, J = 1.4 Hz, OCH$_2$O), 5.72 (m, 7-H), 4.93 (dd, 11-H), 4.18 (m, 3-H), 3.71 (dd, 8-H), 3.64 (dm, 8-H), 3.47 (dd, 10-H), 3.37 (3-OCH$_3$), 3.16 (dd, 10-H), 2.68 (dddd, 4-H), 1.86 (dd, 4-H); position of coupling H-atoms Ha, Hb/J [Hz]: 1.2/10.1; 1.3/2.2; 4$_{ax}$,3/10.6; 4$_{eq}$,2/1.1; 4$_{eq}$,3/5.5; 4$_{eq}$,7/1.1; 4$_{gem}$/11.6; 8$_{gem}$/14.2; 8a,7/3.0; 10$_{ax}$,11/10.0; 10$_{eq}$,11/6.7; 10$_{gem}$/14.4 ^{13}C not reported	*E. caffra*	(38)

24	(+)-15β-D-Glucoerysopine $C_{23}H_{29}NO_8$ (447.49)	150–152 (dark brown solid)	3415, 2920, 1507	279 (3.74), 222 (4.40), 206 (4.38)	6.99 (s, 17-H), 6.75 (s, 14-H), 6.60 (dd, J = 10.1/2.1 Hz, 1-H), 6.02 (d, J = 10.1 Hz, 2-H), 5.79 (br s, 7-H), 4.74 (d, J = 7.2 Hz, 1′-H), 4.00 (m, 3-H), 3.95 (dd, J = 11.0/3.9 Hz, 6′-H$_{ax}$), 3.74 (m, 6′-H$_{eq}$) 3.61 (m, 8-H$_{ax}$), 3.48 (m, 4′-H), 3.48 (m, 2′-H), 3.45 (m, 8-H$_{eq}$), 3.40 (m, 5′-H), 3.40 (m, 10-H$_{ax}$), 3.34 (s, OCH$_3$), 3.30 (m, 3′-H), 2.91 (m, 11-H$_{ax}$), 2.91 (m, 10-H$_{eq}$), 2.67 (m, 11-H$_{eq}$), 2.52 (dd, J = 10.6/5.7 Hz, 4-H$_{ax}$), 1.78 (dd, J = 10.9/10.9 Hz, 4-H$_{eq}$)	E. latissima (39)
				447 (M$^+$, 80), 299 (80), 268 (100), 251 (70)		
		+67.5			147.6 (C-15), 145.8 (C-16), 143.3 (C-6), 133.9 (C-13), 132.2 (C-2), 125.4 (C-12), 124.7 (C-1), 122.6 (C-7), 118.1 (C-17), 114.6 (C-14), 103.8 (C-1′), 77.7 (C-3′), 77.1 (C-3), 77.0 (C-5′), 74.3 (C-2′), 70.8 (C-4′), 67.5 (C-5), 61.9 (C-6′), 57.1 (C-8), 55.9 (OCH$_3$), 44.4 (C-10), 41.3 (C-4), 24.3 (C-11)	
25	(+)-16β-D-Glucoerysopine	158–160	3415, 2920, 1507		7.03 (s, 17-H), 6.79 (s, 14-H), 6.61 (dd, J = 10.1/2.1 Hz, 1-H), 6.04	E. latissima (39)

Table 1 (continued)

Nr.	Trivial name(s)/Formula/Structure	M.p./°C ±[α]$_D$/° cm^2 g^{-1}	IR: $\bar{\nu}$/cm^{-1} UV: λ_{max}/nm (log ε/mol^{-1} dm^3 cm^{-1}) MS: m/z	^{1}H NMR: δ/ppm ^{13}C NMR: δ/ppm	Natural source	Ref.
	C$_{23}$H$_{29}$NO$_8$ (447.49) β-gluc-O HO	(dark brown solid) +76.5	279 (3.74), 222 (4.40), 206 (4.38) 447 (M$^+$, 80), 299 (80), 268 (100), 251 (70)	(d, J = 10.1 Hz, 2-H), 5.79 (br s, 7-H), 4.78 (d, J = 7.2 Hz, 1′-H), 4.00 (m, 3-H), 3.95 (dd, J = 11.0/3.9 Hz, 6′-H$_{ax}$), 3.74 (m, 6′-H$_{eq}$) 3.61 (m, 8-H$_{ax}$), 3.48 (m, 4′-H), 3.48 (m, 2′-H), 3.45 (m, 8-H$_{eq}$), 3.40 (m, 5′-H), 3.40 (m, 10-H$_{ax}$), 3.34 (s, OCH$_3$), 3.30 (m, 3′-H), 2.91 (m, 11-H$_{ax}$), 2.91 (m, 10-H$_{eq}$), 2.67 (m, 11-H$_{eq}$), 2.52 (dd, J = 10.6/5.7 Hz, 4-H$_{ax}$), 1.79 (dd, J = 10.9/10.9 Hz, 4-H$_{eq}$) 145.8 (C-15), 145.1 (C-16), 143.8 (C-6), 134.1 (C-13), 132.1 (C-2), 126.1 (C-12), 125.4 (C-1), 122.8 (C-7), 118.2 (C-17), 113.9 (C-14), 103.6 (C-1′), 77.7 (C-3′), 77.0 (C-3), 77.4 (C-5′), 74.2 (C-2′), 70.7 (C-4′), 67.5 (C-5), 61.8 (C-6′), 57.1 (C-8), 55.9 (OCH$_3$), 44.3 (C-10), 41.2 (C-4), 24.2 (C-11)		
26	Coculidine N-oxide C$_{18}$H$_{23}$NO$_3$ (301.39)	150–152 (HCl: 236–238)	– 287 (3.40), 225 (3.95), 205 (4.37)	spectra not reported	*Cocculus laurifolius*	(40)

| 27 | (+)-8-Oxo-α-erythroidine epoxide

$C_{16}H_{17}NO_5$ (303.31) | —

colourl. oil

+211 | 301 (M^+, 3,4), 285, 284, 283

1715, 1680

250 (sh, 3.69), 216 (4.23)

303 (M^+, 100), 287 (31), 274 (31), 272 (22), 271 (43), 255 (24), 244 (29), 242 (32), 231 (51), 217 (46)

CD (MeOH, c 3.30×10^{-5}): $\Delta\varepsilon$ +6.63 (263), −2.61 (228), +7.87 (208) | = 6.51 (s, 7-H), 6.09 (s, 14-H), 4.51 (dd, $J = 11.7/5.1$ Hz, 17-H_{eq}), 4.36 (m, 10-H_{eq}), 4.17 (dd, $J = 11.7/4.4$ Hz, 17-H_{ax}), 4.05 (d, $J = 3.7$ Hz, 1-H), 3.64 (d, $J = 3.7$ Hz, 2-H), 3.55 (dd, $J = 11.0/5.1$ Hz, 3-H), 3.49 (s, OCH$_3$), 3.08 (ddd, $J = 13.2/13.2/3.7$ Hz, 10-H_{ax}), 2.76 (dd, $J = 13.2/5.1$ Hz, 4-H_{eq}), 2.74 (m, 12-H), 1.94 (m, 11-H_{eq}), 1.73 (dddd, $J = 13.2/12.5/6.6/5.1$ Hz, 11-H_{ax}), 1.55 (dd, $J = 13.2/11.0$ Hz, 4-H_{ax}) | E. poeppigiana (41) |

[a] Inseparable 79:21 mixture with the isomeric known alcohol (+)-erythrinine.

Table 2. New Homoerythrinane Alkaloids

Nr.	Trivial name(s)/ Formula/Structure	M.p./°C ±[α]$_D$/° cm² g^{-1}	IR: $\bar{\nu}$/cm^{-1} UV: λ_{max}/nm (log ε/mol^{-1} dm³ cm^{-1}) MS: m/z	¹H NMR: δ/ppm ¹³C NMR: δ/ppm	Natural source	Ref.
28	1,6α-Epoxyrobustivine C$_{20}$H$_{27}$NO$_5$ (361.44)	no data reported	3575, 3400, 2910, 2830, 1590, 1394, 1305 283 (3.11), 207 (4.55) 361 (M$^+$), 344, 181, 180, 167, 166 (100)	6.55 (15-H), 3.89, 3.83 and 3.79 (3 OCH$_3$), 3.79 (1-H), 3.46 (3-H), 3.46 (12-H), 3.46 (10-H), 3.32 (10-H), 3.02 (8-H), 2.83 (8-H), 2.57 (12-H), 2.50 (4-H$_{eq}$), 2.40 (2-H), 2.28 (7-H), 2.03 (2-H), 1.86 (4-H$_{ax}$), 1.80 (7-H), 1.76 (11-H), 1.58 (11-H) 151.81, 149.90, 141.42, 137.24, 129.44, 108.75 (C-15), 70.57 (C-5), 67.15 (C-6), 64.44 (C-3), 61.34 (OCH$_3$), 60.87 (OCH$_3$), 57.43 (C-1), 56.24 (OCH$_3$), 49.78 (C-10), 45.82 (C-8), 36.05 (C-4), 33.36 (C-2), 27.95 (C-7), 25.72 (C-12), 22.70 (C-11)	*Phelline comosa* Labill. var. *robusta* (Baill.) Loesner	*(42)*
29	18-De-O-methylholidine C$_{19}$H$_{25}$NO$_4$ (331.41)	140 (colourl. crystals) +99	3425, 3100 (s), 1595, 1480, 1440, 1360, 1325 276 (3.09), 208 (4.47); in alcaline solution: 293 (4.18)	6.39 (s, 15-H), 5.50 (m, 1-H), 3.87 (s, OCH$_3$), 3.51 (2m, 12-H$_a$, 10-H$_a$), 3.30 (m, 3-H), 3.25 (s and m, 3-OCH$_3$ and 10-H$_b$), 2.78 (dd and m, 4-H$_{eq}$ and 8-H$_{a,b}$), 2.58 (m, 12-H$_b$ and 2-H$_a$), 2.45 (m, 7-H$_a$), 2.30 (m, 7-H$_b$), 2.02 (m, 2-H$_b$), 1.76 (m, 11-H$_a$), 1.56 (dd and m, 2H, 4-H$_{ax}$)	*Phelline comosa* Labill. var. *robusta* (Baill.) Loesner	*(42)*

		+111	331 (M$^+$), 300, 273, 272, 180 (100)	147.0, 145.3, 141.7, 117.0, 110.8 (CH), 74.2 (C-3), 70.0 (C-5), 60.6 (OCH$_3$), 55.8 (3-OCH$_3$), 50.5 (C-10), 46.8 (C-8), 37.7 (C-4), 32.0 (C-2), 27.5 (C-7), 25.3 (C-12), 23.1 (C-11)	*Cephalotaxus fortunei* (43)
30	Fortunine C$_{19}$H$_{21}$NO$_4$ (327.38)	120–121.5	2980, 2930, 2890, 2820, 1610, 1500, 1480, 1350, 1325, 1298, 1229, 1090, 1030, 915	6.94 (s, 15-H), 6.57 (s, 18-H), 6.01 (br d, $J = 10.2$ Hz, 1-H), 5.85 (d, $J = 1.40$ Hz, OCH$_2$O), 5.75 (dd, $J = 10.2/2.0$ Hz, 2-H), 3.52 (ddd, 3-H), 3.29 (s, OCH$_3$), 3.22/2.62 (m, 11-H$_{a,b}$), 3.20 (m, 7-H), 2.85 (dd, $J = 12.7/4.2$ Hz, 8-H$_{a,b}$), 2.63 (m, 10-H$_{a,b}$), 1.80 (m, 11-H$_{a,b}$), 1.68–1.65 (br d, $J = 10.6$ Hz, 4-H$_{a,b}$)	
		–	327 (M$^+$, 75), 312 (100), 296, 284, 267, 254, 240, 228, 160, 152, 128, 115, 77	146.1 (C-17), 145.1 (C-16), 134.4 (C-13), 133.3 (C-2), 132.8 (C-14), 127.2 (C-1), 111.9 (C-18), 109.5 (C-15), 100.1 (OCH$_2$O), 76.3 (C-3), 69.3 (C-5), 68.9 (C-6), 60.1 (C-7), 56.2 (3-OCH$_3$), 55.5 (C-8), 49.7 (C-10), 35.4 (C-12), 31.7 (C-4), 29.1 (C-11)	
31	Cephalezomine M	colourl. solid	3360, 2930, 1670, 1510, 1200	6.75 (s, 15-H), 6.71 (s, 18-H), 6.05 (s, 1-H), 3.77 (m, 10-H$_b$), 3.51 (d, $J = 14.2$ Hz, 10-H$_a$), 3.35 (s, 8-H$_b$),	*Cephalotaxus harringtonia var. nana* (44)

Table 2 (continued)

Nr.	Trivial name(s)/ Formula/Structure	M.p./°C ±[α]$_D$/° cm^2 g^{-1}	IR: $\bar{\nu}$/cm^{-1} UV: λ_{max}/nm (log ε/mol^{-1} dm^3 cm^{-1}) MS: m/z	^{1}H NMR: δ/ppm ^{13}C NMR: δ/ppm	Natural source	Ref.
	C$_{18}$H$_{23}$NO$_3$ (301.39)	+64	286 (3.26), 233 (3.56), 217 (3.38) 302 (M$^+$ + H)	3.28 (m, 3-H), 3.28 (m, 12-H$_b$), 3.23 (s, OCH$_3$), 3.17 (t, J = 13.0 Hz, 12-H$_a$), 2.84 (m, 4-H$_b$), 2.72 (m, 8-H$_a$), 2.72 (m, 7-H$_{a,b}$), 2.72 (m, 2-H$_b$), 2.08 (m, 2-H$_a$), 2.08 (m, 11-H$_b$), 1.94 (m, 11-H$_a$), 1.77 (t, J = 11.4 Hz, 4-H$_a$) 147.32 (C-17), 144.62 (C-16), 136.88 (C-6), 133.93 (C-13), 126.20 (C-1), 123.41 (C-14), 120.37 (C-18), 120.03 (C-15), 76.79 (C-5), 74.30 (C-3), 56.51 (OCH$_3$), 52.01 (C-10), 49.09 (C-8), 37.46 (C-4), 35.27 (C-12), 32.76 (C-2), 26.82 (C-7), 22.86 (C-11)		
32	2α-Hydroxy-lenticellarine C$_{19}$H$_{25}$NO$_6$ (363.41)	no data reported	3560, 2960, 2950, 2860, 1732, 1710, 1680, 1610, 1510, 1452, 1445, 1370, 1280, 1260, 1207, 1150, 1106, 1080, 933, 750 —	6.88 (s, 15-H), 5.72 (br s, 1-H), 4.43 (dd, J = 3.8 Hz, 2-H), 3.92 (s, 16-OCH$_3$), 3.68 (m, 3-H), 3.47 (s, 17-OCH$_3$), 3.44 (dd, J = 9.15 Hz, 10-H$_b$), 3.34 (s, 3-OCH$_3$), 3.04 (dd, J = 5.15 Hz, 10-H$_a$), 2.96 (m, 8-H$_{a,b}$), 2.62 (t, J = 4.0 Hz, 4-H$_{eq}$), 2.48 (m, 7-H$_{a,b}$), 1.92 (dd, J = 4.12 Hz, 4-H$_{ax}$), 1.62 (br m, 11-H), 1.09 (dd, J = 5.58 Hz, 12-H$_b$), 0.94 (dd, J = 5.9 Hz, 12-H$_a$)	Dysoxylum lenticellare	(45)

33	2α-Methoxy-lenticellarine $C_{20}H_{27}NO_6$ (377.44)	no data reported	363 (M$^+$, 15), 348 (6), 346 (31), 345 (100), 332 (6), 305 (6), 181 (13), 165 (18), 137 (52) 2960, 2940, 2865, 1730 (C=O), 1690, 1630, 1510, 1480, 1440, 1370, 1334, 1280, 1260, 1205, 1150, 1100, 1080, 930, 740 — 377 (M$^+$, 28), 362 (5), 347 (31), 346 (100), 319 (12), 318 (19), 196 (5), 165 (15), 137 (58)	^{13}C not reported 6.90 (s, 15-H), 5.72 (br s, 1-H), 4.42 (d, $J = 3.8$ Hz, 2-H), 3.92 (s, 16-OCH$_3$), 3.82 (s, 2-OCH$_3$), 3.67 (m, 3-H), 3.46 (s, 17-OCH$_3$), 3.44 (dd, $J = 9.15$ Hz, 10-H$_b$), 3.33 (s, 3-OCH$_3$), 3.04 (dd, $J = 5.15$ Hz, 10-H$_a$), 2.96 (m, 8-H$_{a,b}$), 2.62 (dd, $J = 4.12$ Hz, 4-H$_{eq}$), 2.48 (m, 7-H$_{a,b}$), 1.90 (t, $J = 12.0$ Hz, 4-H$_{ax}$), 1.60 (m, 11-H), 1.07 (dd, $J = 5.0/5.8$ Hz, 12-H$_b$), 0.92 (dd, $J = 5.9$ Hz, 12-H$_a$) ^{13}C not reported	*Dysoxylum lenticellare* (45)

similar type. Thus, they mostly are oxygenated at atoms C3, C15, C16 and C3, C16, C17, respectively. Furthermore, there is a small group of C16-deoxygenated erythrinane alkaloids. The saturated parent compounds of both alkaloid series possess the *cis*-configuration of the A/B-moiety independently whether they have been obtained by transformation of natural alkaloids or by synthesis *(27–29)*. The alkaloids are generally dextrorotatory and their absolute configuration at the spiroatom C5 is (*S*) with respect to the basic framework *(17, 22, 24, 30, 31)*. Due to an anomalous substitution pattern there are few exceptions, for instance wilsonine (**227**), with (*R*)-configuration of C5. Atom C3 of erythrinanes is always (*R*)-configured, that of homoerythrinanes, however, is also found exhibiting (*S*)-configuration. For an overview concerning structures, substitution patterns, and properties see *e.g.* refs. *(19)* and *(24)*. Obviously, *Erythrina* alkaloids with *trans*-fused A/B rings do not naturally occur. Nevertheless, the *trans*-erythrinane skeleton has been synthesized by Mondon *(31)* and Desmaële *(32)*.

3. New *Erythrina* Alkaloids

Up to the most recent reviews published in the years 1995/96 nearly 95 erythrinane and 68 schelhammerane type alkaloids were known *(19, 21, 24)*. In this chapter all the new compounds reported in literature from 1995 to 2004 are listed including the natural source, structure, analytical and spectroscopical informations. All in all, there are 12 erythrinane and 6 homoerythrinane type compounds compiled in Tables 1 and 2.

4. Biosynthesis of *Erythrina* Alkaloids

4.1. Erythrinane Alkaloids

The first generally accepted pathway for the biosynthesis of erythrinane alkaloids was established by Barton's group. This route starts with the benzylisoquinoline (*S*)-norprotosinomenine (**34**) as the main precursor, which is cyclized by *para-para* phenol coupling to the neoproaporphine derivative (**35**). This in turn undergoes rearrangement yielding the symmetrical dibenzazonine **36**. Its hydrogenation product **37** is oxidized to the corresponding diphenoquinone (**38**). Finally, intramolecular Michael type addition proceeds to afford erysodienone (**39**) possessing the characteristic erythrinane skeleton *(46–48)*. Both precursors, **34** and **37**, are naturally occurring compounds *(49, 50)* (Scheme 1).

References, pp. 56–62

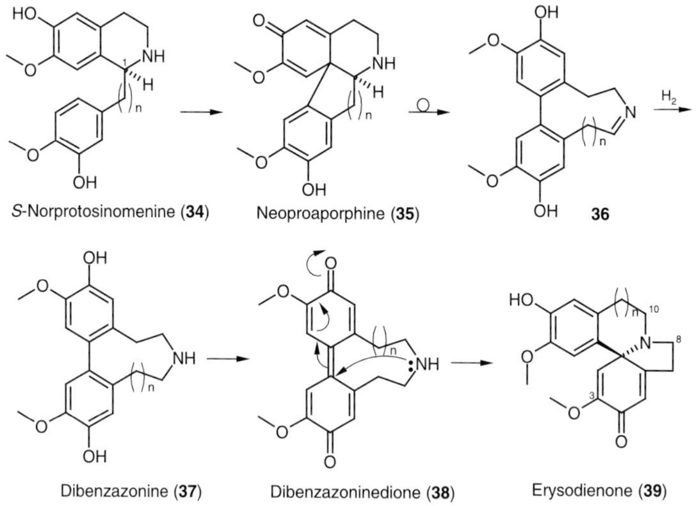

S-Norprotosinomenine (**34**) Neoproaporphine (**35**) **36**

Dibenzazonine (**37**) Dibenzazoninedione (**38**) Erysodienone (**39**)

Scheme 1. Biosynthesis of erythrinane alkaloids according to Barton *et al.* (all formulae: n = 1) (*46–48*)

However, the relative small incorporation rate (0.1–0.25%; (*48*)) of the norprotosinomenine (**34**) into the alkaloid erythraline (**3**) and mainly the fact, that the majority of the important isoquinoline alkaloids, *e.g.* protoberberines, aporphines, bisbenzylisoquinolines, morphinanes, pavines, and benzophenanthridines, are biosynthetically derived from (*S*)-reticuline (**40**: NCH$_3$ instead of NH), have caused a reinvestigation of the biosynthesis pathway outlined in Scheme 1. Thus, in detailed investigations it has been unequivocally shown that (*S*)-norreticuline (**40**) in fact is also the most important biosynthetic precursor for the erythrinane alkaloids, and its incorporation rate exceeds by far that of the isomeric (*S*)-norprotosinomenine (**34**) previously found (7.9% *vs* 0.25% in the case of erythraline). Furthermore, the latter was not converted into any of the alkaloids in the reinvestigation concerned. These results required a new route and mechanism for the biosynthesis of the erythrinane alkaloids depicted in Scheme 2.

According to model reactions in this field previously carried out (*51*), the initial *para-para* coupling of (*S*)-norreticuline (**40**) should led – differently from the previous route – to the morphinandienone derivative norisosalutaridine (**41**) rather than to the neoproaporphine derivative (**35**). The latter, after generation of the benzo[1,3]dioxole function giving noramurine (**42**), can rearrange via the neospirinic ion (**43**) to the unsymmetrical dibenzazonine (**44**). Oxidation at the free phenolic unit

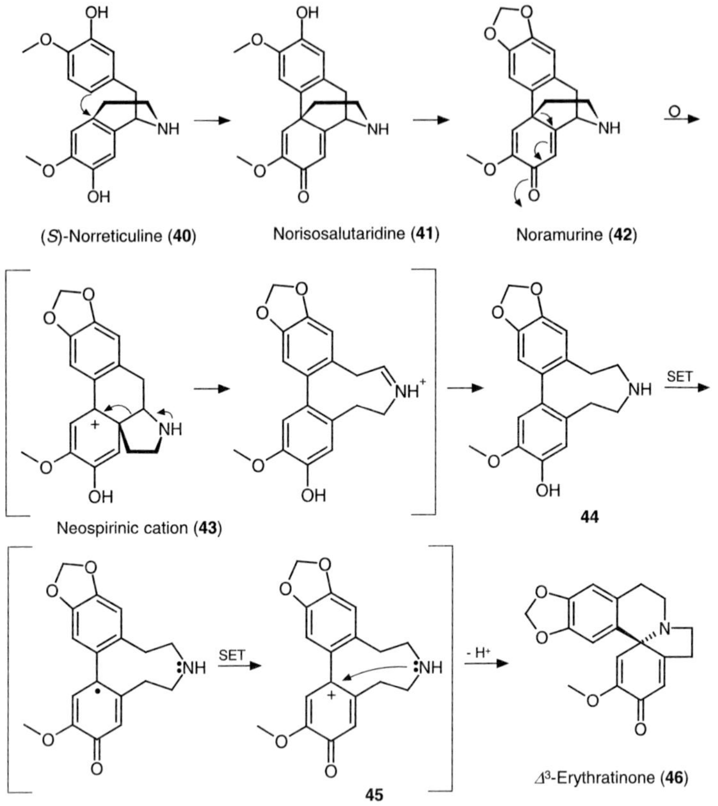

Scheme 2. New biosynthetic sequence of erythrinane alkaloids proposed by Zenk *et al.* (*52*)

proceeding through a SET mechanism can afford the diallylic cation (**45**), which is assumed to react with the nitrogen atom to generate Δ^3-erythratinone (**46**). A symmetrical intermediate of the diphenoquinone type (**38**) postulated previously could be excluded based on feeding experiments with ^{13}C-labelled precursors. Thus, *e.g.* using (*S*)-[1-^{13}C]-norreticuline a ^{13}C-enrichment should occur at C10 as well as at C8 of the isolated erythrinane alkaloid **39**. However, the enrichment has been observed exclusively at C10 (*52*). The subsequent steps leading to erythraline (**3**) are similar to those already established (*53*).

4.2. Homoerythrinane Alkaloids

The biosynthesis of homoerythrinane alkaloids has been proposed to proceed by the same pathway as that of the erythrinanes according to

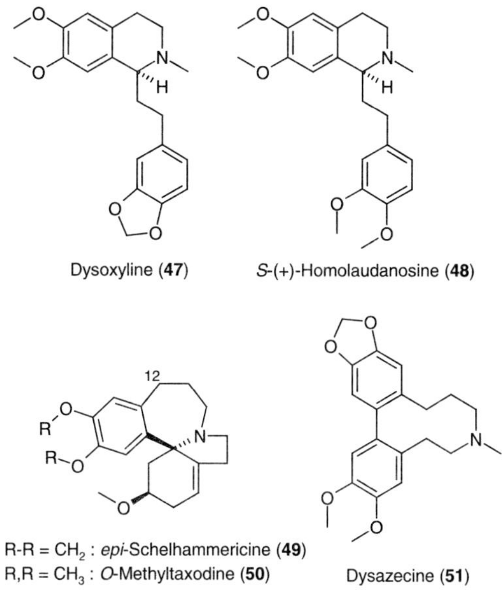

Dysoxyline (**47**) S-(+)-Homolaudanosine (**48**)

R-R = CH$_2$: *epi*-Schelhammericine (**49**)
R,R = CH$_3$: *O*-Methyltaxodine (**50**)   Dysazecine (**51**)

Fig. 3. Phenethylisoquinolines **47** and **48** isolated from plants, which produce homoerythrinane and dibenzazecine alkaloids **49**, **50**, and **51**

Scheme 1 starting from the homologous S(+)-1-phenethyltetrahydroisoquinoline precursor (**34**, n = 2) via the corresponding dibenzazecine (**37**, n = 2). Supports for these assumptions are based on the isolation of several naturally occurring phenethylisoquinolines, *e.g.* dysoxyline (**47**) and S(+)-homolaudanosine (**48**) from plants that produce the homoerythrinane alkaloids 3-epi-schelhammericine (**49**), *O*-methyltaxodine (**50**), and especially the dibenzazecine alkaloid dysazecine (**51**) possessing the same skeleton as the postulated biosynthetic intermediate (**37**, n = 2) (*22*) as shown in Fig. 3.

But until now there is no experimental evidence for this route concerned, which possibly – with regard to that of the erythrinane alkaloids – will demand a revision.

5. Syntheses of *Erythrina* Alkaloids and *Eyrthrina* Type Compounds

Because of their unique structures and their biological activities there has been much interest in the total synthesis of *Erythrina* alkaloids. Thus numerous synthesis approaches to the tetracyclic framework, mainly to that of the erythrinanes, have been developed until now. Contrarily, only few routes have been reported for the synthesis of the ring C-homologous schelhammeranes.

5.1. Methodical Classification

A practical classification of all syntheses especially concerning the erythrinane ring system is based on the idea which of the three alicyclic rings A, B, or C is completed in the final step. Thus in principle two alternatives are conceivable:

- The completion of one of the rings mentioned (in the following called "route A, B, or C")

or

- the sequential or simultaneous formation of more than one cycle e.g. A/B, A/C, B/C in the final or A/B/C in one step, correspondingly called "route A/B, B/C" etc. (Schemes 3 and 4).

Depending on the bond(s) to be formed several subtypes result from both cases as outlined in Schemes 3 and 4: e.g. ring C formation can be achieved using method (a) or method (b) [=route C(a) or C(b)]. Accordingly, two different pathways are available for the simultaneous construction of rings B and C [=route B/C(a) and B/C(b)].

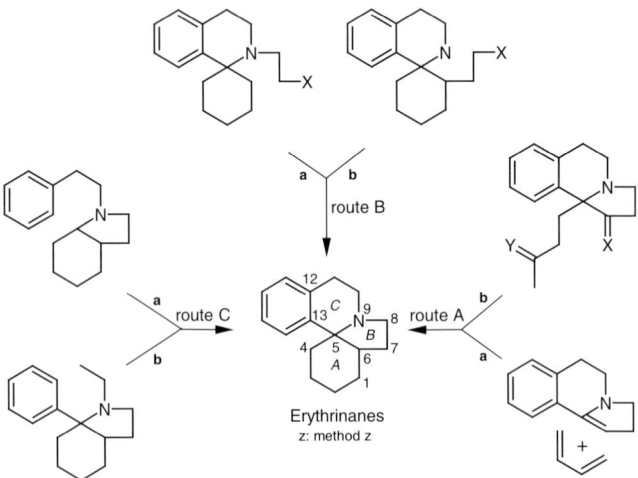

Scheme 3. Strategies for the synthesis of aromatic erythrinane alkaloids (5.2.1.): Generation of one alicyclic ring in the final step

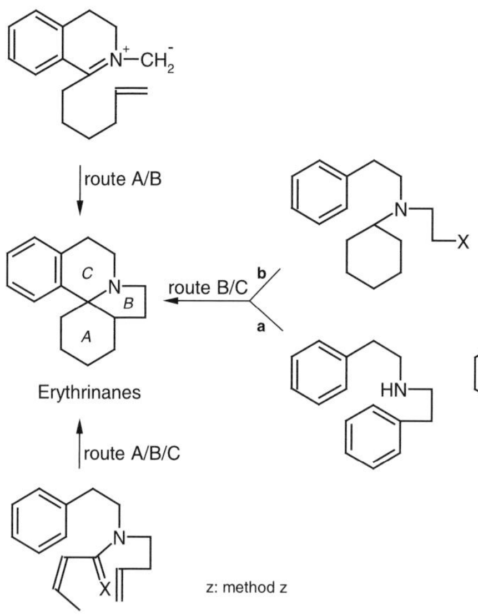

Scheme 4. Strategies for the synthesis of aromatic erythrinane alkaloids (5.2.2.): Generation of more than one alicyclic ring by sequential or tandem cyclization in the final or in one step

In this classification C-homoerythrinane syntheses remain out of consideration, since the construction of their framework is far more difficult than that of the erythrinanes. Nevertheless, several useful approaches to this alkaloid group are included in this section (see below).

Concerning the non-aromatic erythrinane alkaloids, only one single total synthesis has been achieved until now. Thus, the synthesis of (±)-cocculolidine (**15**, Fig. 2) has been completed in 0.42% overall yield through 21 steps (*54*). Furthermore, several synthesis approaches to ring D oxaerythrinane frameworks are known (*19*).

5.2. Erythrinanes

5.2.1. Final Formation of One Ring

5.2.1.1. Ring C (Route C)

5.2.1.1.1. Cyclization of N-Phenethylhydroindole Derivatives (Route C(a))

Generation of a C5 quaternary center – frequently in form of an acyliminium ion – and acid catalyzed reaction with the aromatic atom C13 – related to the Pictet-Spengler reaction – is a widespread method to

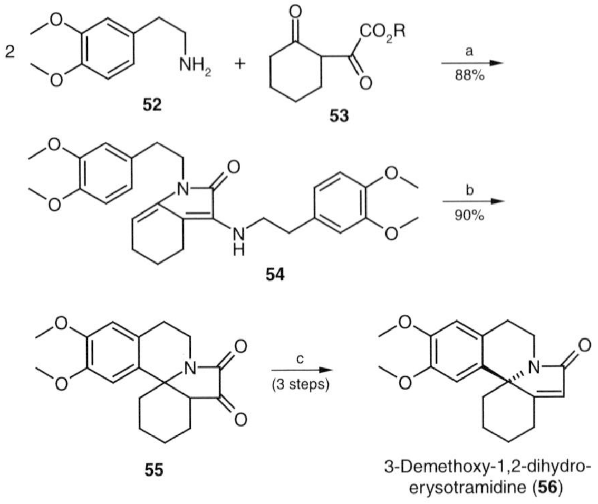

Scheme 5. Synthesis of 3-demethoxy-1,2-dihydroerysotramidine (**56**) (*56, 57*): a) 180°C in xylene/N$_2$; b) dil. H$_3$PO$_4$/CH$_3$OH/Δ; c) *i*: H$_2$/Raney-Ni/NaOH in EtOH; *ii*: *p*-TsCl in pyr/0°C; *iii*: refl. collidine

synthesize the erythrinane skeleton. This approach has often been realized with the simultaneous formation of ring B (s. below, route B/C(a)), especially in the early investigations (*20, 55*).

A clean C(a) route represents one of Mondon's syntheses starting from two equivalents homoveratrylamine (**52**) and cyclohexanoyl glyoxylic acid (**53**) to afford the *N*-phenethylhydroindole derivative (**54**) in high yield. Intramolecular cyclization with phosphoric acid gives the 7,8-dioxoerythrinane (**55**), which can be transformed to 3-demethoxy-1,2-dihydroerysotrine (**56**) (*56, 57*).

A related general pathway (Scheme 6), based on a subsequent introduction of the α-dicarbonyl moiety starts from the enamines **59** generated from homoveratrylamine (**52**) and 2-oxocyclohexane carboxylic derivatives **58**. They are reacted with oxalylchloride yielding also hydroindole precursors **60** (X=CO), which on immediate treatment with a Lewis acid or after reduction to the carbinol (X=CHOH) have been cyclized to the corresponding highly functionalized erythrinanes **61** in excellent yields (*58–60*). From a chiral phenethylamine, *e.g.* the (*S*)-(+)-dimethoxyphenylalanine methylester (**57**) the hydroindole derivatives **60** (R^1 = CO$_2$CH$_3$) have been diastereoselectively prepared by a similar sequence (*de* = 50–60%). After elevating the diastereomeric excess to *de* = 82% by a kinetic resolution, enantiotype erythrinane alkaloids such as (−)-3-demethoxyerythratidinone (**62**) are available (*60*).

References, pp. 56–62

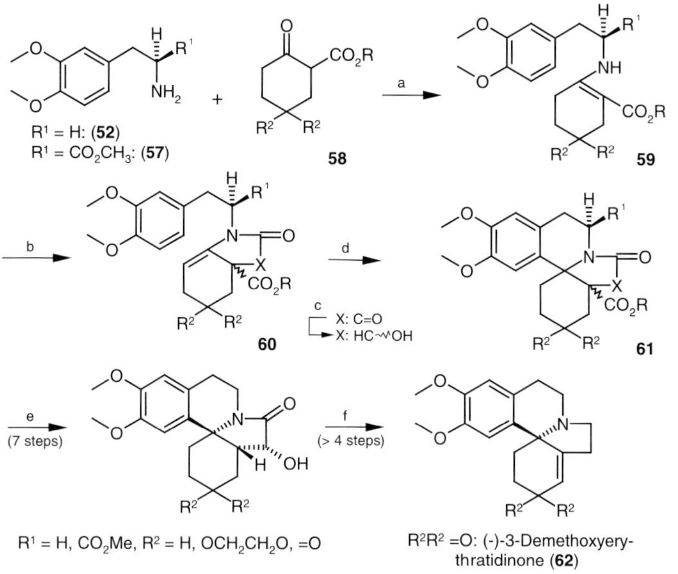

Scheme 6. Synthesis of (−)-3-demethoxyerythratidinone (**62**) (*58–60*): a) neat at 110°C or in EtOH at 100–150°C/sealed tube; b) (COCl)$_2$ in benzene or Et$_2$O/0 to −15°C; c) NaBH$_4$ in EtOH/0°C; d) BF$_3$·Et$_2$O in CH$_2$Cl$_2$/20°C or refl.; e) 7 steps and f) 4 steps (*60*)

A more efficient asymmetric approach to the erythrinane core has been achieved by utilizing a bicyclic lactam template of Meyers. In the present case condensation of racemic cyclohexanoylacetic acid (**64**) with the chiral benzylaminoethanol **63** stereoselectively gives the required tricyclic lactam **65**, which on treatment with titanium tetrachloride has been cyclized – via the *N*-acyliminium intermediate **66** – to the desired erythrinane derivative **67** with 98% yield. Finally, the hydroxymethyl auxiliary group at the 10-position has been readily removed by an established three-step procedure affording the chiral (−)-3-demethoxy-tetrahydroerysotramidine (**68**) (*61*) (Scheme 7).

An interesting pathway concerning the generation as well as the reactivity of the required incipient *N*-acyliminium ion **71a** represents the NBS-promoted cyclization of the phenethylhydroindolinone precursor **71** (Scheme 8). This reaction markedly depends on the polarity of the solvent and only proceeds to give the desired cyclization product **71b** when acetonitrile has been used; in methylene chloride or tetrahydrofuran the cyclization of **71 → 71b** does not occur. Precursor **71** is available from homoveratrylamine (**52**) and the phenylsulfanyl cyclohexanoyl acetic acid ester **69** via the angularly substituted bicyclic lactam **70**. Finally, **71b**

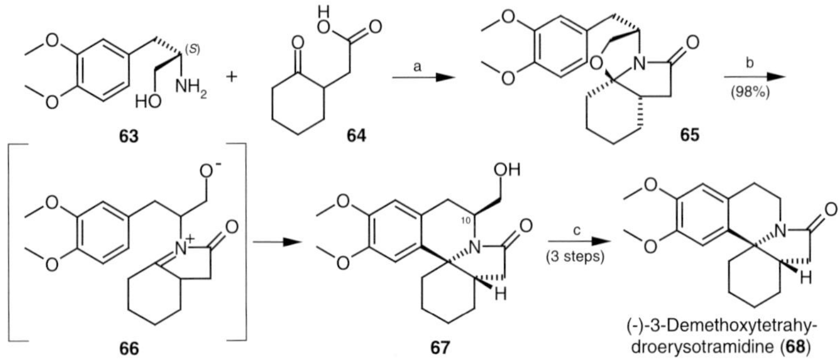

Scheme 7. Synthesis of (−)-3-demethoxytetrahydroerysotramidine (**68**) (*61*): a) toluene/Δ; b) TiCl$_4$ in CH$_2$Cl$_2$/−78°C; c) *i*: Dess-Martin periodane in CH$_2$Cl$_2$; *ii*: Rh(PPh$_3$)$_2$(CO)Cl/ 1,3-bis(diphenylphosphino)propane/xylene/Δ; *iii*: H$_2$/Pd–C, EtOH

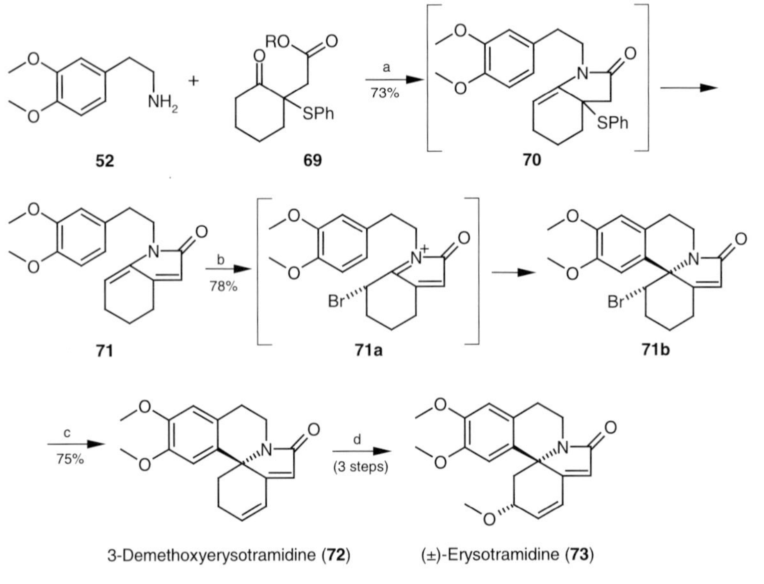

Scheme 8. Synthesis of (±)-erysotramidine (**73**) (*62*): a) TFA/xylene/160°C; b) NBS in CH$_3$CN; c) DBU/refluxing xylene; d) *i*: SeO$_2$/HCO$_2$H; *ii*: CH$_3$COCl/EtOH; *iii*: CH$_3$I/ KOH in THF

has been dehydrohalogenated yielding 3-demethoxyerysotramidine (**72**), which affords (±)-erysotramidine (**73**) in three additional steps (*62*).

Suitably functionalized hydroindole precursors have been efficiently generated from alkenyltrichloroacetamides **76**, which undergo radical cyclization promoted by nickel powder providing the hydroindolone **77**

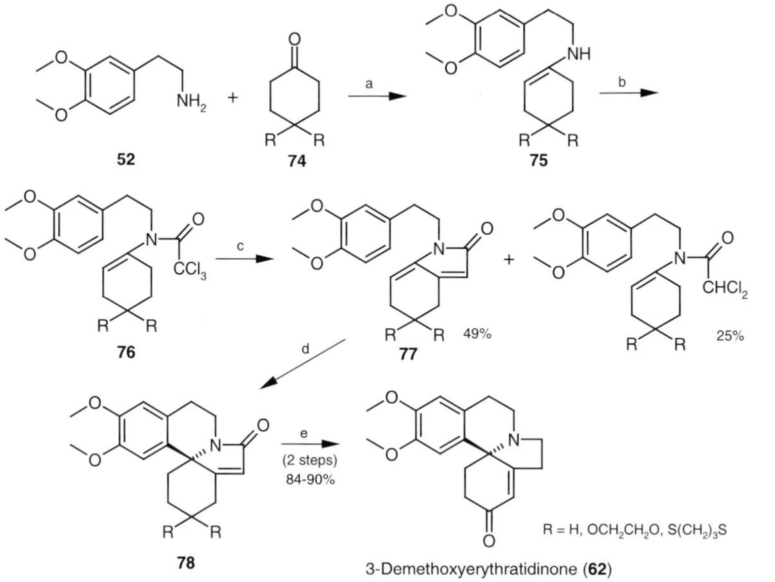

Scheme 9. Synthesis of 3-demethoxyerythratidinone (**62**) (*63*): a) in toluene, Dean-Stark; b) Cl$_3$CCOCl/Et$_3$N in toluene, 0°C; c) Ni/AcOH in 2-propanol/refl.; d) *p*-TsOH in benzene/refl.; e) R = S(CH$_2$)$_3$S, *i*: LiAlH$_4$/AlCl$_3$ in THF/Et$_2$O, 0°C; *ii*: *N*-chlorosuccinimide/AgNO$_3$ in CH$_3$CN/H$_2$O

(Scheme 9). On treatment with *p*-toluene sulfonic acid in refluxing benzene they afford the 8-oxo-$\Delta^{6,7}$-erythrinane **78** with at least 84% yield. Further two steps involving the reduction of the amide moiety and deprotection of the ketone with concomitant migration of the double bond lead to 3-demethoxyerythratidinone (**62**) (*63*). The required educts **76** are readily available by reaction of homoveratrylamine (**52**) and cyclohexanones **74** via enamine **75** followed by N-acylation with trichloroacetylchloride.

Besides this widespread pathway based on the intermediate *N*-acyliminium ion, several other established methods have been applied to construct the C5/C13 bond of the erythrinane skeleton. Thus, the Heck reaction has proved to be an attractive approach to the target compounds. The synthesis reported by Rigby (Scheme 10) starts with a smooth [1 + 4] cycloaddition of certain isocyanides to the vinyl isocyanate **79** affording the required hydroindolone **80**. Then the iodoarene moiety has been installed by *N*-alkylation with the phenethylmesylat **81** giving the N-alkylated precursor **82**. Cyclization of **82** under Heck conditions yields the expected 7,8-dioxoerythrinane **83** as a single diastereomer, which then has been converted to (±)-2-*epi*-erythrinitol (**84**) in twelve additional steps (*64*).

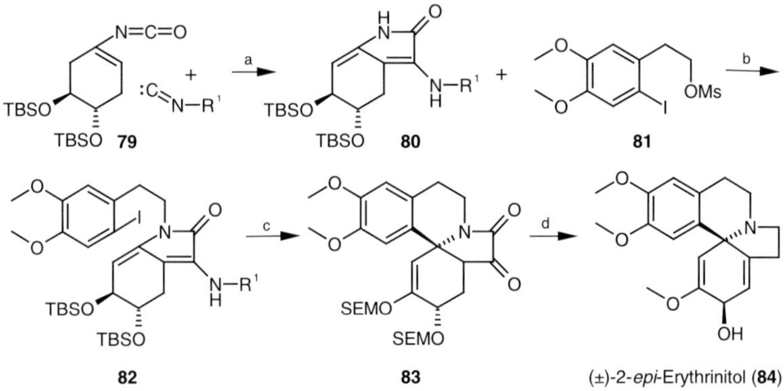

Scheme 10. Synthesis of (±)-2-*epi*-erythrinitol (**84**) (*64*): a) R^1: *c*-hexyl, in CH$_3$CN/ 20°C; b) NaH in DMF; c) *i*: TBAF in THF; *ii*: SEMCl/*i*-Pr$_2$NEt; *iii*: Pd(OAc)$_2$/(*o*-tol)$_3$P, TEA in CH$_3$CN/H$_2$O; d) 12 steps, overall yield 7.2% (*64*)

A new diastereoselective route to aromatic *cis*-erythrinanes represents the combined intramolecular Strecker and Bruylants reactions of the phenethyl-cyclohexanylethylamines **88** (Scheme 11). Deprotection of the carbonyl function and addition of potassium cyanide causes the Strecker reaction to give the angularly substituted hydroindole derivatives **89** in nearly quantitative yields. Then the Bruylants reaction is

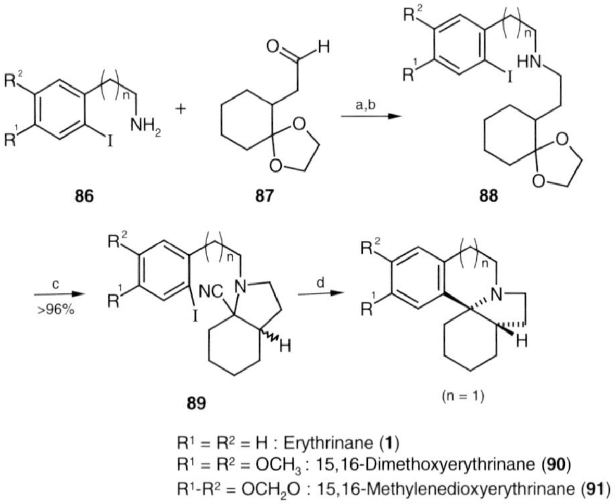

R^1 = R^2 = H : Erythrinane (**1**)
R^1 = R^2 = OCH$_3$: 15,16-Dimethoxyerythrinane (**90**)
R^1-R^2 = OCH$_2$O : 15,16-Methylenedioxyerythrinane (**91**)

Scheme 11. Synthesis of erythrinanes (**1**, **90**, **91**; n = 1) (*29*): a) equimol. educts in toluene/ Dean-Stark; b) NaBH$_4$ in CH$_3$OH/refl.; c) in 2 molequiv. 2 *N* HCl/H$_2$O, 60°C, then addition of 2 molequiv. KCN/H$_2$O at 20°C; d) 1.05 molequiv. 2 *M* *i*-PrMgCl in THF at −50° to +60°C

started with *i*-propylmagnesium chloride providing the parent tetracycles **1**, **90**, and **91** with 70–80% yields. Contrarily, the corresponding C-ring homologue schelhammeranes (*e.g.* **2**) are not available on this route. The required educts **88** have been easily prepared by reductive N-alkylation of iodophenethylamines **86** with cyclohexanylacetaldehyde (**87**) (*29*).

5.2.1.1.2. Cyclization of Angulary Arylated Hydroindole Derivatives (Route C(b))

Alternatively, C-ring formation has been achieved by connecting the atoms C11 and C12 of the erythrinane core. This route principally involves the generation of angularly arylated hydroindoles, their N-alkylation with an appropriate C_2-unit followed by the final cyclization step. Thus the hydroindole **93**, accessible from 5-(nitromethyl)-1,3-benzodioxole (**91**) and the oxoheptenoic acid ester **92**, has been reacted in a hetero Michael addition with phenyl vinylsulfoxide providing the required N-substituted precursor **94**. The latter in turn has been cyclized in a Pummerer reaction to

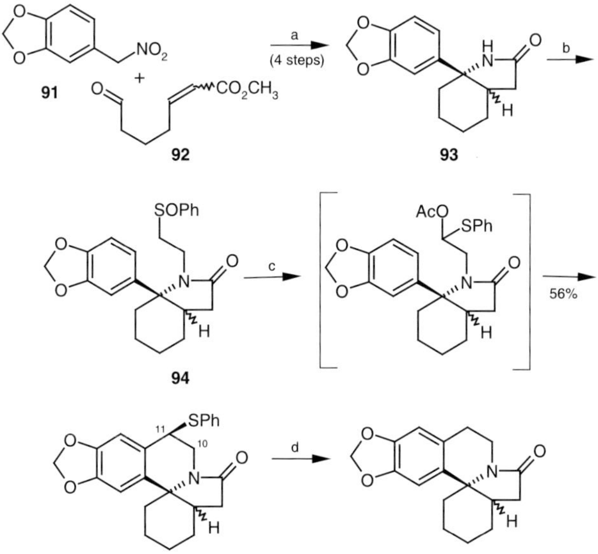

cis/trans-15,16-methylenedioxy-8-oxoerythrinane (**96**)

Scheme 12. Synthesis of *cis*- and *trans*-15,16-methylenedioxy-8-oxo-erythrinane (**96**) (*32*): a) *i*: Amberlyst A-21 in Et_2O; *ii*: $EtO_2CN^-SO_2N^+Et_3$ in benzene/Δ; *iii*: $NaBH_3CN$ in $AcOH/CH_3OH$; *iv*: H_2/Raney-Ni in CH_3OH, then refl. in toluene; b) $CH_2{=}CH\text{-}SOPh$/ $[(CH_3)_3Si]_2NNa$ in THF/−78° to 20°C; c) Ac_2O/refl., then $SnCl_4$ in CH_2Cl_2; d) Bu_3SnH/ AIBN in toluene/Δ afford a *cis/trans* mixture of the 11-thiophenylether **95** as main product together with some sulfur-free 10,11-dehydrogenated material. Desulfuration with tri-*n*-butyl-stannane/azodiisobutyronitrile (AIBN) leads to 15,16-methylenedioxy-8-oxoerythrinane (**96**). Starting from the diastereomeric pure hydroindoles **93** this sequence is useful to prepare both stereomers of the erythrinane **96** (*32*) (Scheme 12).

Similar phenylhydroindole precursors have been prepared more efficiently by a strategy (Scheme 13), which generally allows the construction of the A ring of the erythrinane as well as of the homoerythrinane core. The basic feature of this pathway involves the Diels-Alder addition, or alternatively, the [2 + 2] photocycloaddition of functionalized 1,3-butadienes to yield suitably substituted phenylpyrroles. This reaction followed by 1,3-anionic rearrangement preforms at least the A/B-moiety of the alkaloid frameworks. In the actual case the [2 + 2] cycloaddition of the diene **98** to the dioxophenylpyrrole **97** possessing a sulfanylethyl

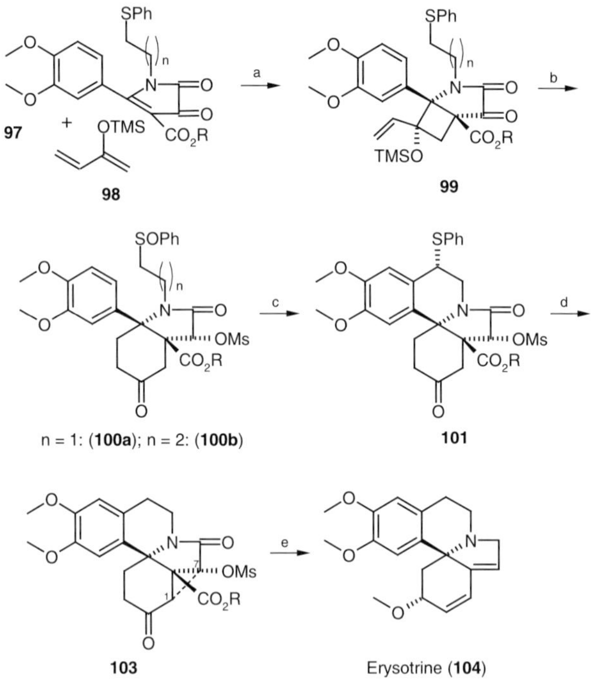

Scheme 13. General synthesis of erythrinanes (**104**) (*65*): a) $h\nu$ in DME/0°C; b) *i*: NaBH$_4$ in CH$_3$OH/−30°C; *ii*: TBAF in THF/−30°C; *iii*: MsCl in pyr/20°C, then NaIO$_4$ in H$_2$O/CH$_3$OH/CH$_2$Cl$_2$, 0°C; c) TFAA in CH$_2$Cl$_2$/20°C; d) *i*: Bu$_3$SnH/AIBN in toluene/Δ; *ii*: DBU in toluene/Δ; e) according to Ref. (*76*)

unit at the nitrogen atom, affords the vinylcyclobutane derivative **99** in high yield. The 1,3-rearrangement followed by transformations of the sulfanyl- and the C3-carbonyl groups provide the phenylhyroindoles **100**, which in turn have been cyclized – analogously to Scheme 12 – by Pummerer reaction giving the erythrinane derivative **101**. This can be converted to the 1,7-cycloerythrinane derivative **103** which allows to generate the alkaloid erysotrine (**104**) (*65*). The same route is applicable to prepare the corresponding homoerythrinane alkaloids (see Scheme 37).

Both, the photo- and the Diels-Alder cycloadditions have been also extended to pyrroloisoquinolines representing one of the most efficient strategies for the construction of the erythrinane as well as the homoerythrinane cores according to route A(a) (see below *e.g.* Schemes 19 and 39).

An new alternative route for the synthesis of erythrinanes using the Friedel-Crafts acylation for C-ring completion is outlined in Scheme 14. The arylhydroindoles **107** ($m=2$, $n=1$), diastereoselectively prepared from the corresponding cycloalkanoylalkylamines **106** by Strecker and subsequent Bruylants reaction ($m=2$; $n=1$), have been N-alkylated with bromoacetic acid ester yielding the required precursor **108**, which on treatment with trifluoromethanesulfonic acid cyclizes to a separable mixture of the *cis*-11-oxoerythrinanes **109** and **110**. Realkylation of the phenolic group in **109** with diazomethane affords the methoxyketones **110** in low to moderate total yields. Contrarily, an attempted cyclization of the precursor lacking an activating *para* substituent in the aromatic

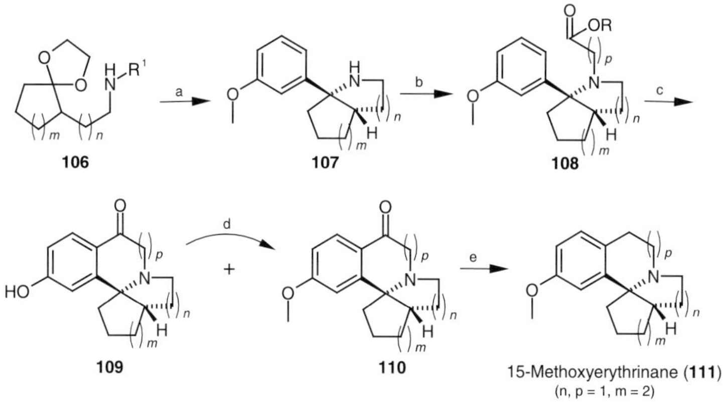

Scheme 14. Synthesis of 15-methoxyerythrinane (**111**) (*28*): a) *i*: $2N$ HCl/H$_2$O, then KCN/H$_2$O/20°C; *ii*: ArMgX/THF, then toluene/Δ; b) BrCH$_2$CO$_2$Et/Na$_2$CO$_3$ in CHCl$_3$/Δ; c) F$_3$CSO$_3$H/110°C; d) excess CH$_2$N$_2$ in Et$_2$O/CH$_3$OH/20°C; e) H$_2$/Pd in Ac$_2$O/HClO$_4$, 60°C, $6.5 \cdot 10^6$ Pa

unit failed. The same is true for educts possessing dihydroxylated aromatic moieties, *e.g.* dimethoxy or methylenedioxy functions. Finally, the carbonyl group is catalytically removed to obtain the parent 15-methoxyerythrinane (**111**). The sequence can be extended to the synthesis of related ring homologue frameworks, *e.g.* to schelhammerane (see Scheme 38), B-homoerythrinane (m, n = 2; p = 1), A-norerythrinane (m, n, p = 1), and to the hitherto unknown A-norschelhammerane (m, n = 1; p = 2) using the corresponding educts as shown in Scheme 14 (*28*).

5.2.1.2. Formation of Ring B (Route B)

Various strategies have been examined to complete the erythrinane framework by B ring generation in the last step, starting from appropriate C5-spiroisoquinolines as depicted in Scheme 3 [route B(a) or B(b)].

5.2.1.2.1. Cyclization of N-Substituted C5-Spiroisoquinoline Derivatives (Route B(a))

The spiroamine **113** conveniently available by a Pictet-Spengler like cyclization of the enamine **112** (*66*) has been alternatively converted to

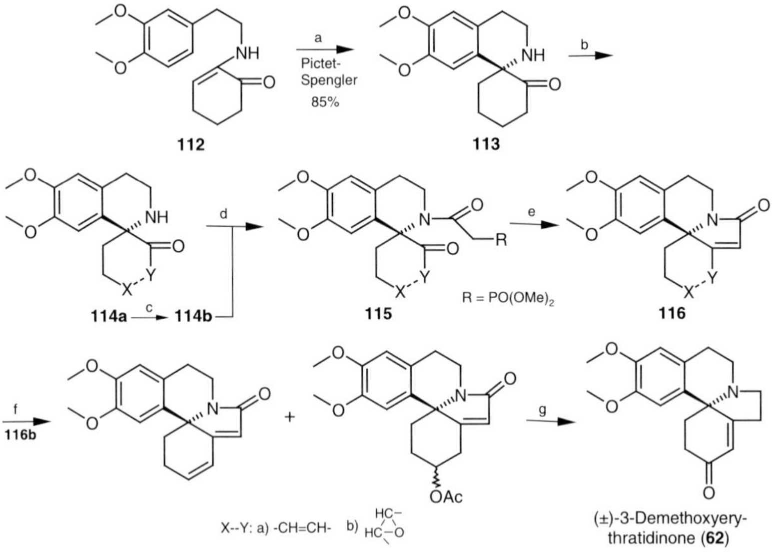

Scheme 15. Synthesis of (±)-3-demethoxyerythratidinone (**62**) (*67*): a) TFA/Δ; b) *i*: (tBuO)$_2$CO in CHCl$_3$/refl.; *ii*: LDA/PhSeCl in THF; *iii*: NaIO$_4$ in EtOH; *iv*: TFA in CH$_2$Cl$_2$; c) H$_2$O$_2$/NaOH in CH$_3$OH; d) (MeO)$_2$P(O)–CH$_2$COCl/pyr in CH$_2$Cl$_2$; e) aqueous KOH/benzene/20°C; f) Zn/AcOH/Ac$_2$O; g) *i*: AlH$_3$ in THF; *ii*: Swern oxidation

References, pp. 56–62

the corresponding α,β-unsaturated ketone **114a** or by an additional step to the epoxide **114b** followed by N-acylation with dimethyl phosphonoacetyl chloride providing the phosphonates **115a** and **115b**. Intramolecular Wadsworth-Emmons reaction smoothly gives the erythrinanes derivatives **116a** and **116b** in good yields. The latter has been transformed to (±)-demethoxyerythratidinone (**62**) by several further steps involving the reductive cleavage of the epoxide moiety, the LiAlH$_4$ reduction of the amide unit, and Swern oxidation of the hydroxy function with concomitant migration of the double bond (*67*) (Scheme 15).

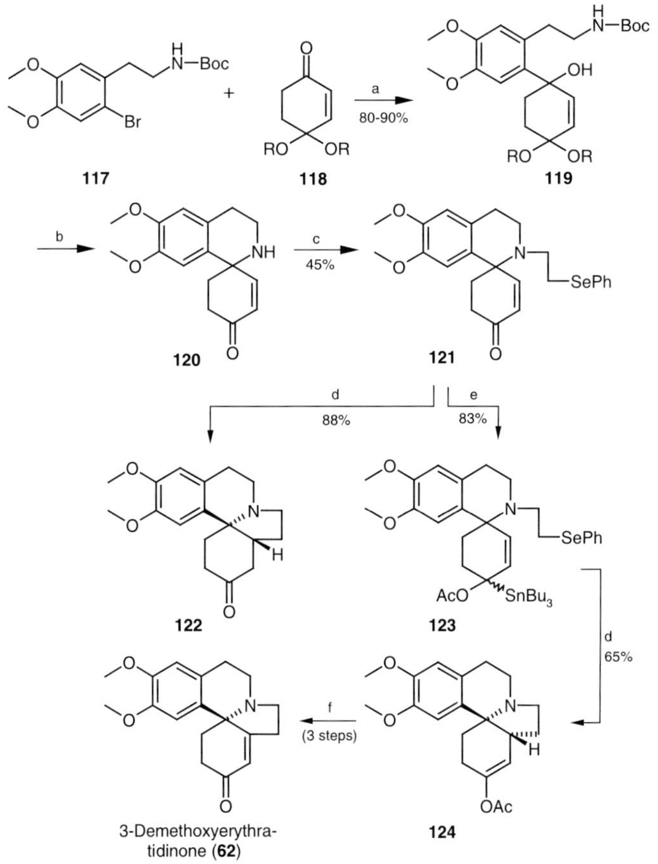

Scheme 16. Synthesis of (±)-demethoxyerythratidinone (**62**) (*68*): a) KH/THF/25°C, then *n*-BuLi in THF/−78°C; b) *i*: (CH$_3$)$_3$SiOTf in CH$_2$Cl$_2$/−78°C; *ii*: DBU; c) PhSeCH$_2$CHO/NaBH$_3$CN in THF/MeOH; d) Bu$_3$SnH/AIBN in benzene; e) *i*: Bu$_3$SnLi in Et$_2$O/−78°C; *ii*: Ac$_2$O/Et$_3$N/DMPA in CH$_2$Cl$_2$; f) *i*: CH$_3$Li in THF; *ii*: PhSeCl/−78°C; *iii*: NaIO$_4$ in THF/H$_2$O

Starting with a spiroaminoketone **120** related to **113** (Scheme 15) the B ring construction has been performed via a Michael like free radical cyclization route (Scheme 16). Thus, **120** has been reductively alkylated with phenylselenoacetaldehyde affording the N-alkylated precursor **121**, which gives in the presence of tri-*n*-butylstannane/AIBN the 2-oxoerythrinane **122** in high yield. Contrarily, treatment of **121** with tri-*n*-butyllithium-thiostannane followed by acetanhydride leads to an 1:1 mixture of the diastereomeric acetoxystannanes **123**, which have been radically cyclized under the same conditions yielding the erythrinane enolester **124**. This can be converted to 3-demethoxyerythratidinone (**62**) in a three step sequence (*68*). The educt **120** has been readily prepared by transformation of the teriary alkohol **119**, which is available from the N-protected 2-bromohomoveratrylamine **117** and the enone ketal **118** (*69*).

Another alternative of this pathway offers the use of the C5-spiroisoquinoline **126** accessible from the dihydroisoquinolinium salt **125** by photocyclization (Scheme 17). Thus, the B ring can be completed by intramolecular Claisen condensation as well as by intramolecular C-alkylation giving the 1,7-dioxo- and the 1-oxoerythrinanes **127** and **129**. Removal of the CO-group via the $\Delta^{1,2}$-derivative **130** affords the *cis*-dimethoxyerythrinane (**90**) (*70*).

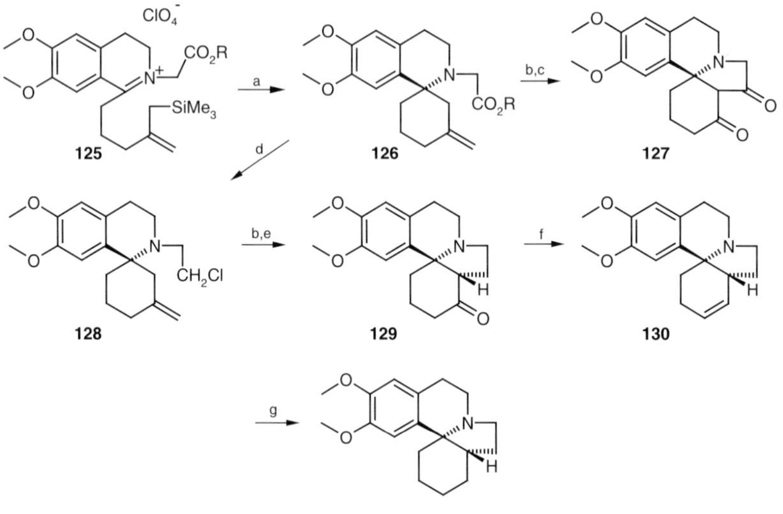

cis-15,16-Dimethoxyerythrinane (**90**)

Scheme 17. Synthesis of 15,16-dimethoxyerythrinane (**90**) (*70*): a) $h\nu$/CH$_3$OH; b) OsO$_4$/NaIO$_4$ in dioxane/H$_2$O; c) NaOEt in EtOH/0°C; d) *i*: LiAlH$_4$ in Et$_2$O/0–25°C; *ii*: MsCl/Et$_3$N in Et$_2$O/25°C; e) DBU in THF/60°C; f) *i*: LiAlH$_4$ in THF/0° to 25°C; *ii*: MsCl/Et$_3$N in Et$_2$O/THF; *iii*: DBU in THF/80°C; g) H$_2$/Pd in THF/25°C/10^5 Pa

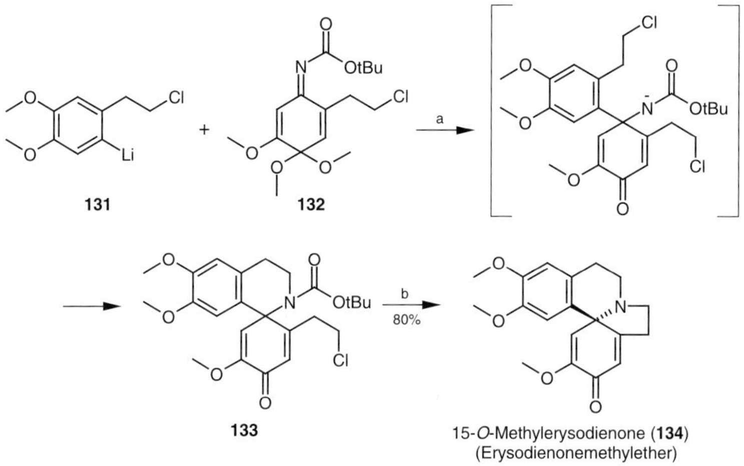

Scheme 18. Synthesis of 15-*O*-methylerysodienone (**134**) (*71*): a) −78°C to reflux temp.; b) F$_3$CCO$_2$H/*p*-Ts OH/20°C, then KOH in H$_2$O/THF

5.2.1.2.2. Cyclization of C6-Substituted C5-Spiroisoquinoline Derivatives (Route B(b))

The final B ring formation by intramolecular N-alkylation requires a C$_2$ side chain attached to the A ring of the C5-spiroisoquinoline educt, *e.g.* **133**. This has been similarly obtained as the related educt **119** (Scheme 16) by reaction of the lithiated phenethyl derivative **131** with the N-protected quinone imide ketal **132**. After deprotection the crude amine undergoes cyclization to form the 15-*O*-methylerysodienone (**134**) in 80% yield (*71*) (Scheme 18).

5.2.1.3. Formation of Ring A (Route A)

5.2.1.3.1. Cycloaddition to Pyrroloisoquinolines (Route A(a))

Formation of the A ring to complete the tetracyclic erythrinane core has been achieved starting from appropriate pyrroloisoquinolines. The most important routes are based – as already mentioned above (Scheme 13) – on the Diels-Alder reaction, or alternatively, on a [2 + 2] photocycloaddition between O-functionalized 1,3-butadiene derivatives, preferably 1,3-bis (trimethylsilyloxy)butadiene or 1-methoxy-3-trimethylsilyloxybutadiene (**136**), and dioxopyrroloisoquinolines (**137a**) (Scheme 19). Thus, the Diels-Alder method regio- and stereoselectively gives the angularly

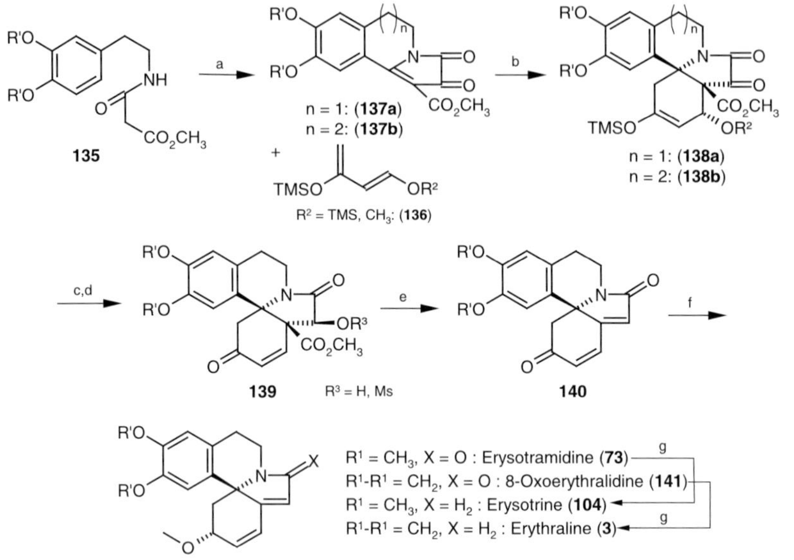

Scheme 19. Synthesis of dienoid type erythrinanes (**3**, **73**, **104**, **141**) (*72*): a) *i*: PPE in CHCl$_3$/reflux; *ii*: (COCl)$_2$ in Et$_2$O/0°C; b) sealed tube/Δ; c) LiBH$_4$ in THF/−60°C, then HCl/THF/reflux; d) MsCl/pyr, 20°C; e) MgCl$_2$ in DMSO/Δ, sealed tube; f) *i*: Al(*i*-PrO)$_3$ in *i*-PrOH/reflux, PTLC of 3-OH-isomers; *ii*: KOH/Et$_4$NBr/CH$_3$I in THF, 20°C; g) AlH$_3$ in THF/20°C

6-carboxylated erythrinanediones **138a** in high yields. These are closely related to those obtained according to route C(a), *e.g.* **61** (Scheme 6) and can be preferably transformed via the intermediates **139** and **140** to the 1,6-dienoid type erythrinanes **3**, **73**, **104**, and **141**. The educts **137a** are readily prepared by the reaction of a phenethylamine with chloroformylacetat providing the amides **135** followed by sequential Bischler-Napieralski reaction and oxalylation (*72*).

Starting from 3,4-dimethoxyphenyl-L-alanine ester (**142**) the sequence of Scheme 20 has been proved to be suitable for the preparation of enantiomeric products via **144** → **145** → **140**, the latter represents a valuable educt to natural alkaloids, *e.g.* (+)-erysotrine (**104**) or (+)-erysotramidine (**73**) (*73*, *74*).

On the other hand, applying the [2+2] photocycloaddition method already described in Scheme 13, the pyrroloisoquinoline **137a** provides the related highly functionalized 2,8-dioxoerythrinane (**147a**) through the analogue cyclobutane intermediate **146**. Further transformation of **147a** leads to erysotramidine (**73**) and erysotrine (**104**) (*75*, *76*) as shown

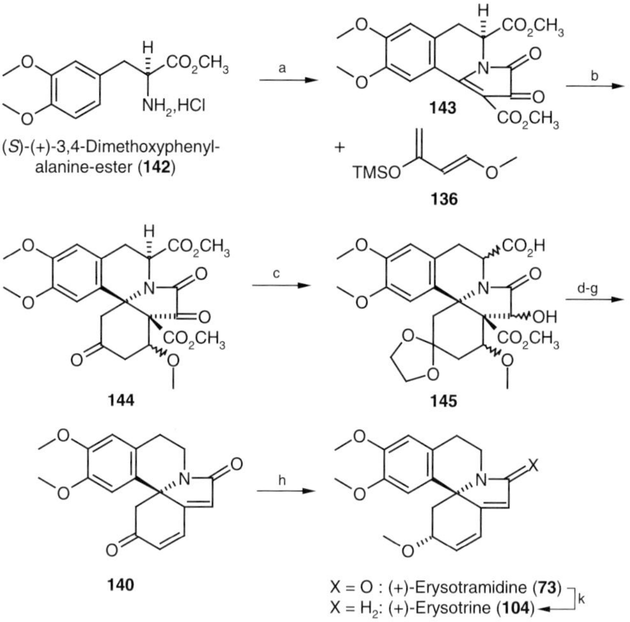

Scheme 20. Chiral synthesis of erythrinanes (**73, 104**) (*73, 74*): a) *i*: ClCOCH$_2$CO$_2$CH$_3$/ K$_2$CO$_3$ in Et$_2$O/H$_2$O, −15° to 10°C; *ii*: PPE/100°C; *iii*: (COCl)$_2$ in Et$_2$O/CH$_2$Cl$_2$/0°C; b) 109 Pa in CH$_2$Cl$_2$/20°C, then treatment with SiO$_2$; c) *i*: ethyleneglycol/*p*-TsOH in benzene/reflux; *ii*: NaBH$_4$ in CH$_3$OH/0°C; *iii*: NaOH in CH$_3$OH/20°C; d) *i*: *N*-methylmorpholine, *i*-Bu-chloroformate in THF/−10°C; *ii*: *N*-hydroxypyridinethione-Na/ Et$_3$N in THF/−10°C; *iii*: hν/*t*-BuSH in THF/0°C; e) HCl in THF/80°C, then PTLC of 7-OH-stereomers; f) MsCl/pyr; g) CaCl$_2$ in DMSO/Δ; h) *i*: NaBH$_4$/CeCl$_3$ in CH$_3$OH/20°C; *ii*: CH$_3$I/KOH/Et$_4$NBr in THF/20°C; k) AlH$_3$/AlCl$_3$ in EtOH/THF/0°C

in Scheme 21. In a similar manner the corresponding schelhammerane intermediates **147b** are accessible, when starting from pyrrolobenzazepines **137b** instead of the pyrroloisoquinolines, which open an approach to natural alkaloids (*cf.* Scheme 39 below) (*77, 78*).

5.2.1.3.2. *Intramolecular Aldol Condensation of Angularly Substituted Pyrroloisoquinoline (Route A(b))*

From the pyrroloisoquinoline carboxylic acid ester **149** (Scheme 22) generated from homoveratrylamine (**52**) and dioxopentenoic acid ester (**148**) by a combined Michael addition/condensation, the A ring has been constructed by an intramolecular aldol condensation of the diketone **150** affording 3-demethoxyerythratidinone (**62**) (*79*).

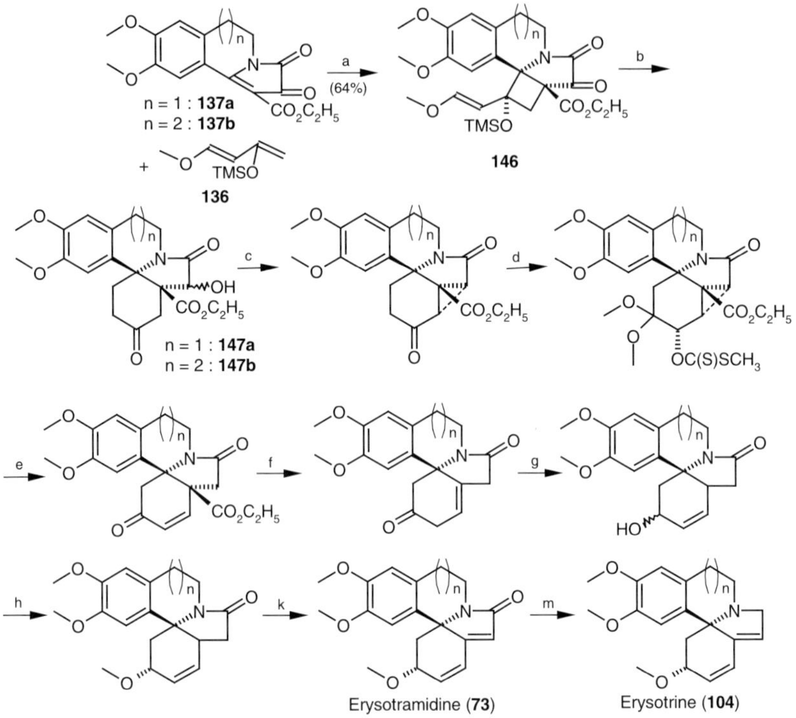

38   E. Reimann

Scheme 21. Synthesis of erythrinanes (n = 1) (**73, 104**) (*75, 76*): a) *hν* in acetone/0°C; b) *i*: NaBH$_4$ in EtOH/0°C; *ii*: TBAF in THF/−30°C; *iii*: H$_2$/Pd in acetone; c) *i*: MsCl in pyr/20°C; *ii*: DBU in toluene/reflux; d) *i*: PhSeCl/Et$_2$O · BF$_3$ in THF/reflux, then Hg(ClO$_4$)$_2$ in CH$_3$OH/0°C; *ii*: NaBH$_4$ in CH$_3$OH, then HCl; *iii*: NaH/imidazole/CS$_2$ in THF/reflux; e) Bu$_3$SnH/AIBN in toluene/reflux, then HCl in acetone/50°C; f) CaCl$_2$/3-ethylpentane-3-thiol in DMSO/Δ, sealed tube; g) *i*: DBU in benzene/Δ; *ii*: NaBH$_4$/CeCl$_3$ in CH$_3$OH/0°C, MPCL of 3-OH stereomers; h) *i*: NaH/imidazole/Bu$_4$NHSO$_4$/CH$_3$I/70°C; k) LDA/PhSeCl in THF, then NaIO$_4$ in CH$_3$OH/H$_2$O; m) AlH$_3$ in Et$_2$O/THF/20°C

5.2.2. Simultaneous Formation of More Than One Ring

Tandem and multicascade processes, more accurately also called domino reactions, belong to a growing group of reactions, which often allow an efficient regio- and stereoselective approach to complex compounds and ring systems in a single operation (*80*). This strategy also has been successfully applicated to construct the alicyclic rings A, B, and C of the erythrinane framework. Strategies reported in this field can be classified as depicted in Scheme 3. The final ring formation widely utilized in former as well as in recent investigations has been conducted

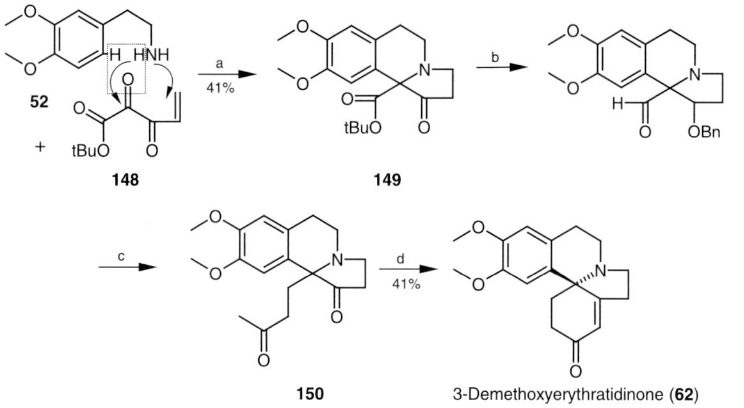

Scheme 22. Synthesis of 3-demethoxyerythratidinone (**62**) (*79*): a) in CH$_2$Cl$_2$, then POCl$_3$; b) *i*: NaBH$_3$CN in AcOH/0° to 20°C; *ii*: NaH/BnBr in THF; *iii*: LiAlH$_4$ in Et$_2$O/0°C; c) *i*: (MeO)$_2$P(O)CH$_2$COCH$_3$/NaH in toluene/reflux; *ii*: H$_2$/Pd in acetone/ 3.8·10^5 Pa, then H$_2$/Pd/catal. HCl in acetone; *iii*: NaBH$_4$ in CH$_3$OH/0°C; *iv*: (COCl)$_2$/ DMSO in CH$_2$Cl$_2$/−78°C, then Et$_3$N/20°C; d) NaOH in CH$_3$OH/reflux

following route B/C. In contrast, only few syntheses have been accomplished hitherto based on the routes A/B and A/B/C.

5.2.2.1. Simultaneous Formation of Rings B and C (Route B/C)

5.2.2.1.1. Cyclization of Secondary Diphenethyl- or (Cycloalkyl)ethyl-phenethylamine Derivatives (Route B/C(a))

The first syntheses of the erythrinane framework reported by Belleau (*81*, *82*) and Mondon (*83–86*) have been performed using cyclization of the isomeric ketalamides **151** with polyphosphoric acid via the assumed

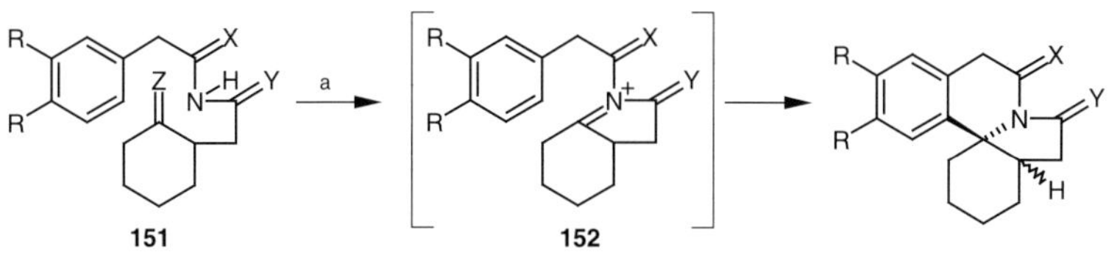

R = H; X = O; Y = H$_2$; Z = O: 10-Oxoerythrinane (**153**)
R = OCH$_3$; X = H$_2$; Y = O; Z = (R'O)$_2$: Demethoxytetrahydroerysotramidine (**68**)
R = OCH$_2$O; X = H$_2$; Y = O; Z = (R'O)$_2$: 15,16-Methylenedioxy-8-oxoerythrinane (**96**)

Scheme 23. Synthesis of 8- and 10-oxoerythrinanes (**68**, **96**, **153**) (*83–86*): a) P$_4$O$_{10}$/PPA or PPA in varying concentrations, 20° to 100°C

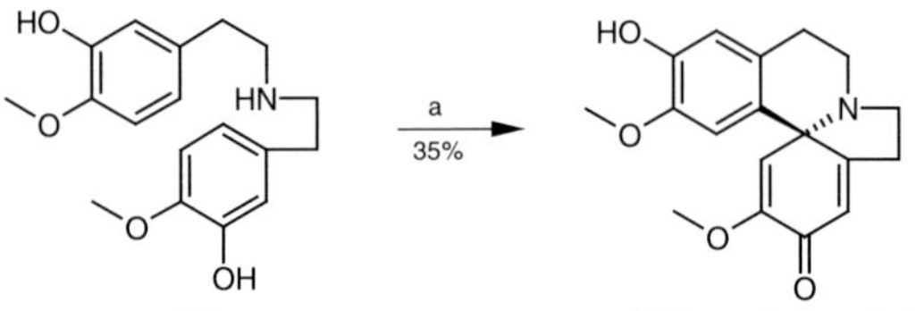

154 (±)-Erysodienone (**39**)

Scheme 24. Synthesis of (±)-erysodienone (**39**) (*87, 88*): a) $K_3[Fe(CN)_6]/Na_2CO_3$ in H_2O/CH_2Cl_2

intermediate *N*-acyliminium ion (**152**) giving the oxoerythrinanes **68**, **96**, and **153** (Scheme 23).

The same type of simultaneous C5/C13 and C5/C9 bond formations has been effected in several biomimetic syntheses starting from symmetrical phenolic bases **154**, which on oxidative coupling provide the expected (±)-erysodienone (**39**) (*87, 88*) as shown in Scheme 24.

In a further biomimetic approach (Scheme 25), the phenolic educt **157**, related to **154** of Scheme 24, and conveniently available from the corresponding phenylacetic acid **155** and the phenethylamine **156**, first has been photochemically converted to the dibenzazocine **158** followed by oxidative cyclization to the corresponding 3-demethoxy-Δ^3-erythratidinone (**159**). This in turn can be selectively hydrogenated yielding 3-demethoxyerythratidinone (**62**) (*89*).

In this connection nonoxidative cyclization strategies are of general interest. They start also from diphenethylamines and involve a Birch

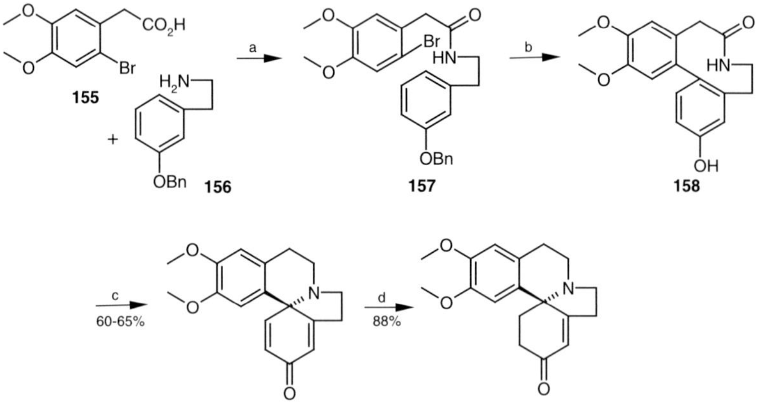

155 + 156 → 157 → 158 → 159 → 3-Demethoxyerythratidinone (**62**)

Scheme 25. Synthesis of 3-demethoxyerythratidinone (**62**) (*89*): a) in decaline/reflux; b) *i*: $h\nu$ in $NaOH/CH_3OH$; *ii*: $NaBH_4/Et_2O \cdot BF_3$ in THF; c) PbO_2 in benzene or $K_3[Fe(CN)_6]/NaHCO_3$ in $CHCl_3/H_2O$; d) H_2/Pd in EtOH/20°C, 10^5 Pa

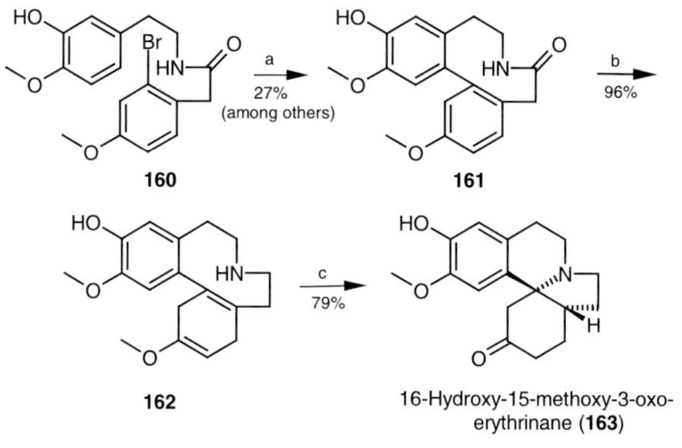

Scheme 26. Synthesis of 16-hydroxy-15-methoxy-3-oxoerythrinane (**163**) (*90*, *91*): a) $h\nu$ in $CH_3OH/NaOH$; b) *i*: $NaBH_4/Et_2O \cdot BF_3$; *ii*: Na/NH_3 in $CH_3OH/Et_2O/THF/-60°$ to $70°C$; c) 10% H_2SO_4 in $DMF/60°C$

reduction as a main feature. Thus, the dibenzazocine **161**, obtained by photocyclization of **160**, gives the Birch 1,4-diene product **162**, which has been cyclized with diluted sulfuric acid affording the 16-hydroxy-15-methoxy-3-oxo-erythrinane **163** (Scheme 26).

On the other hand, the related educt **164** has been immediately reduced leading to the corresponding 1,4-cyclohexadiene **165**, which on treatment with acid provides the 2,8-dioxo-16-hydroxy-15-methoxyerythrinane (**166**) in 90% yield (*90*, *91*) (Scheme 27).

In a new attractive B/C(a) route the required N-disubstituted amine derivatives, *e.g.* the metalated carboxamides **169**, have been generated *in situ*, reacting at first the primary phenethylamins **52** or **167** with trimethylaluminum followed by the enolacetates of cyclohexanoyl carboxylic acids **168**. The intermediates **169** eliminate acetic acid affording the *N*-acyliminium ions (**170**), which cyclize to the desired erythrinanes **68** and **96**. Due to the typical ^{1}H NMR shift of 14-H given ($\delta = 6.93$ ppm (*31*)) the products should possess the B/C *cis*-configura-

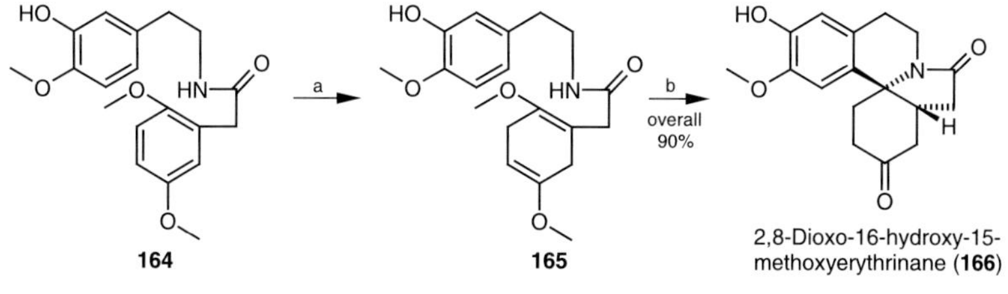

Scheme 27. Synthesis of 2,8-dioxo-16-hydroxy-15-methoxyerythrinane (**166**) (*90*, *91*): a) Birch reduction; b) 10% H_2SO_4 in DMF

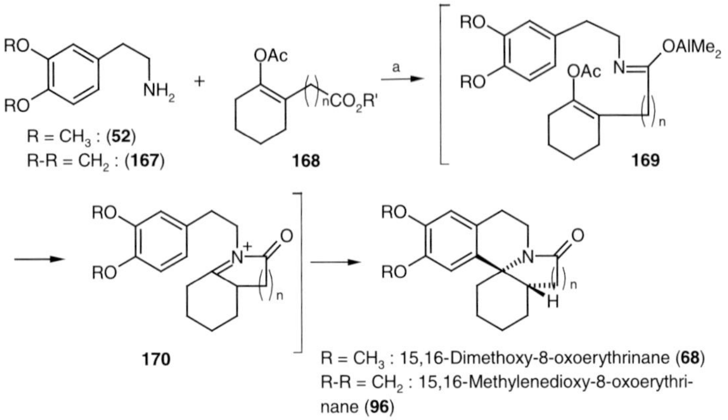

Scheme 28. Synthesis of 15,16-dioxygenated-8-oxo-erythrinanes (**68**, **96**; n = 1) (*92*): a) 2 equiv. of Al(CH$_3$)$_3$ in benzene/Δ

tion. This pathway of Scheme 28 opens an efficient approach to the erythrinane as well as to the B-homoerythrinane core (n = 1 or 2) in a single step (*92*).

5.2.2.1.2. Cyclization of Tertiary Cyclohexyl-ethyl-phenethyl-amide Derivatives (Route B/C(b))

Domino sequences especially induced by the Pummerer reaction are well known to be very useful for the construction of complex polycyclic

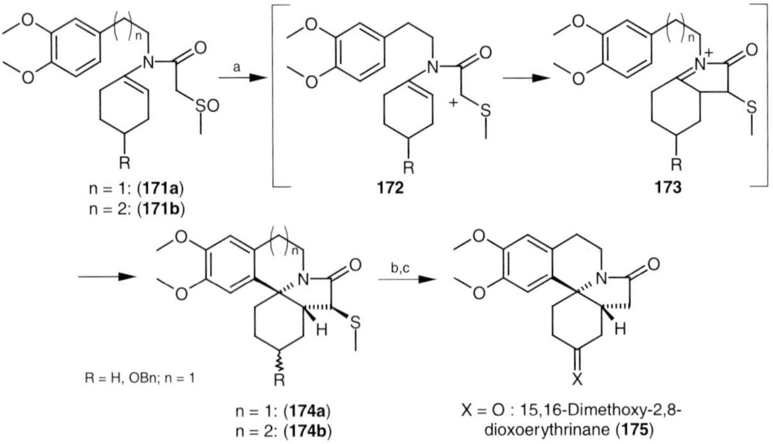

Scheme 29. Synthesis of oxoerythrinanes (**175**) (*94*): a) *p*-TsOH in CH$_2$Cl$_2$/reflux; b) Raney-Ni in EtOH; c) CrO$_3$ in pyr/CH$_2$Cl$_2$ (R = OBn)

References, pp. 56–62

compounds (*cf. e.g.* (*93*)). They also have been successfully applied to the construction of the erythrinane core starting from relative simple building blocks. For instance, in a one step synthesis the sulfoxide **171a** (Scheme 29) undergoes sequential B/C ring cyclization via thionium/acyliminium intermediates **172** and **173** yielding 50–60% of the erythrinane derivative **174a**, which can be transformed to the 2,8-dioxoderivative **175**. 15,16-Dimethoxy-8-oxo-erythrinane (X=H$_2$; **68**) is available using the same route (*94*).

Since thionium ions, *e.g.* **177** also can be generated by treatment of thioacetals or thioketals with dimethyl(methylthio) sulfonium tetrafluoroborate (DMTSF), a similar cyclization sequence occurred when the amido thioketal **176** was used as an educt (Scheme 30). Starting from the educt **176** the ring formation according to route B/C(a) takes place giving 71% of 15,16-dimethoxy-8-oxoerythrinane (**68**) in one step (*95*).

Furthermore, sulfur containing compounds, especially α-S-functionalized acetamide derivatives are suitable precursors to assemble the erythrinane framework via a combined radical/acyliminium ion cyclization. Thus, treatment of the xanthate **178** (Scheme 31) with lauroyl peroxide causes B ring generation via the radicals **179** and **180**. After further oxidation of the latter forming the acyliminium ion **181**, catalytical amounts of *p*-toluenesulfonic acid induces the ring closure to the aromatic unit furnishing 15,16-dimethoxy-2,8-dioxoerythrinane (**175**) in 82% yield (*96*).

In the same manner the radical B ring formation also can be induced by manganese(III) acetate in the presence of copper(II) acetate or copper(II) triflate. Thus, the α-(methylthio) acetamide precursor **182**

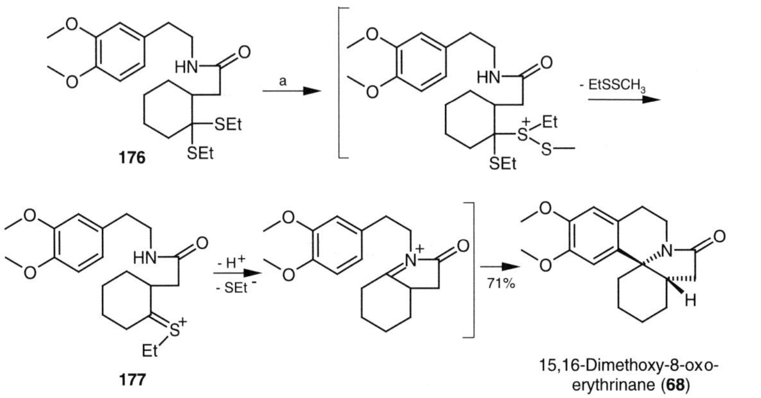

Scheme 30. Synthesis of 15,16-dimethoxy-8-oxoerythrinane (**68**) (*95*): a) [(CH$_3$)$_2$S$^+$SCH$_3$]BF$_4^-$ (DMTSF) in CH$_2$Cl$_2$/reflux

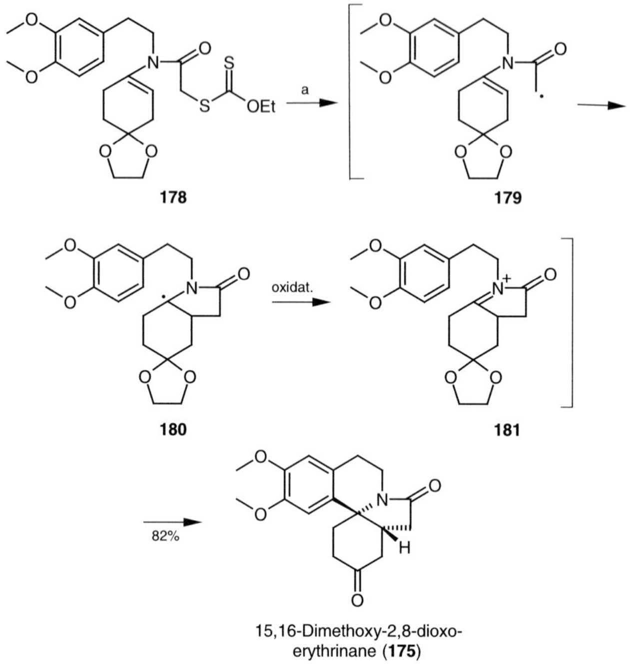

Scheme 31. Synthesis of 15,16-dimethoxy-2,8-dioxo-erythrinane (**175**) (*96*): a) lauroylperoxide in CH$_2$Cl$_2$/reflux, then catalytic amounts of *p*-TsOH/reflux

(Scheme 32) provides in only one step the erythrinane **183** (*97*), which allows the access to the naturally occurring (±)-3-demethoxyerythratidinone (**62**) (*98*).

5.2.2.2. Cyclization of N-Substituted 1-Acyldihydroisoquinolinium Derivatives (Route A/B)

Until now only one example is reported concerning a simultaneous A/B-ring construction for the synthesis of the erythrinane skeleton. The required key intermediate 1-acyldihydroisoquinoline **186** (Scheme 33) has been conveniently prepared by reaction of homoveratryl isonitrile **184** with appropriate carboxylic acid halogenides, as *e.g.* 5-hexenoyl chloride (**185**). On treatment of **186** with trimethylsilylmethyltriflate followed by cesium fluoride the intermediately generated azomethinylide **187** undergoes [3 + 2] cycloaddition affording the 4-oxoerythrinane **188** in 70% yield. In the same manner precursors involving an acetylenic dipolarophile, *e.g.* the 1-hexynoyldihydroisoquinoline **189** provide the corresponding $\Delta^{6,7}$-oxoerythrinane **190** in 42% yield (*99*).

References, pp. 56–62

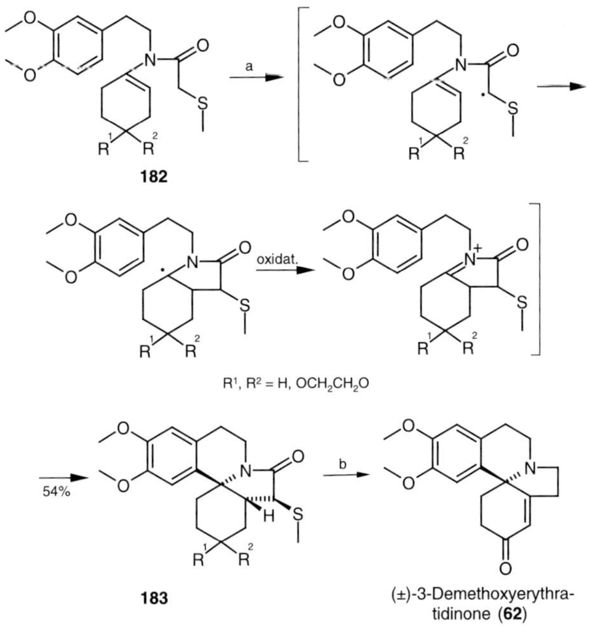

Scheme 32. Synthesis of (±)-3-demethoxyerythratidinone (**62**) (*97, 98*): a) 6–10 equiv. Mn(OAc)$_3$, 1 equiv. Co(OTf)$_2$ in F$_3$CCH$_2$OH/reflux; b) i: NaIO$_4$ in H$_2$O/acetone/0°C, then NaHCO$_3$/toluene/reflux; *ii*: LiAlH$_4$/AlCl$_3$ in Et$_2$O/THF, −15°C; *iii*: HCl in acetone/reflux

5.2.2.3. Cyclization of a Highly Functionalized Homoveratrylimide (Route A/B/C)

Finally, there is also only a single report describing the sequential formation of the rings A, B, and C of the erythrinane framework in one step. Starting from the complex homoveratrylimide derivative **191** this triple cascade process involves – apart from the initial Pummerer reaction **191** → **192** – the Diels-Alder reaction **192** → **193** as well as the final acyliminium ion cyclization **194** → **195** providing the erythrinane **195** in 83% yield. This in turn could be converted to (±)-erysotramidine (**73**) by a sequence already reported (*76*) (Scheme 34). The requisite educt **191** has been smoothly prepared through six steps in 45% overall yield (*80*).

5.3. Homoerythrinanes

The construction of the C-homoerythrinane framework is far more difficult than that of the erythrinane, that is, methods developed for

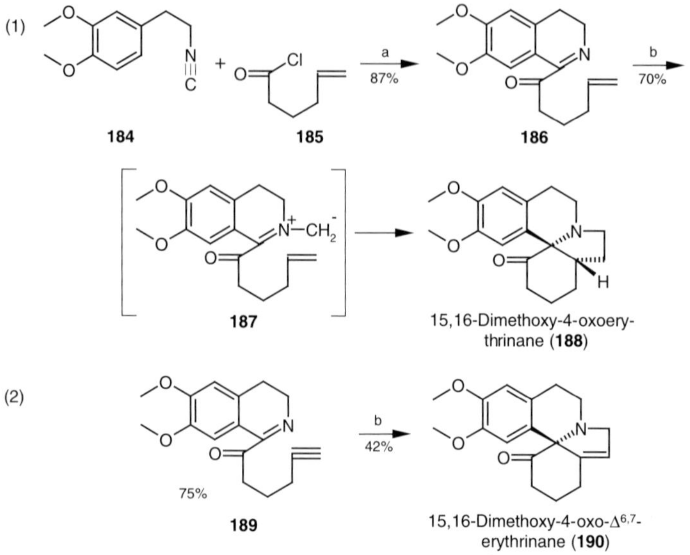

Scheme 33. Synthesis of 4-oxoerythrinanes (**188**, **190**) (*99*): a) in CH$_2$Cl$_2$/25°C, then AgOTf in CH$_2$Cl$_2$/−20°C; b) (CH$_3$)$_3$SiCH$_2$OTf in CH$_2$Cl$_2$/25°C, then CsF in 1,2-DME/65°C

the synthesis of erythrinanes are – apart from very few exceptions (s. below) – not automatically transferable to schelhammerane synthesis. For instance, the parallel approaches using several homologue educts or intermediates of the erythrinane synthesis, *e.g.* the *N*-arylpropyl-enamides **171b** as well as the corresponding dioxopyrrolobenzazepine **137b** failed to give the target compounds **174b** or **138b** at all or afford them only in less than 5% yield on acid-mediated cyclization or on Diels-Alder reaction with trimethylsilyloxybutadiene (*cf.* Schemes 19 and 29) (*59, 100, 101*). This substantial difference concerning the synthesis of both ring systems is attributed to the skewed nature of the sevenmembered nucleus. In this connection MMX force field calculations (PC Model (*102*)) are of interest revealing the parent *cis*-erythrinane to be essentially more stable than the corresponding C-homologue schelhammerane. Assuming the chair form of the benzocycloheptene core (*103*) and the same relative configuration (5*S*,6*S*)/(5*R*,6*R*) comparable to that of the erythrinane skeleton (*cf.* Fig. 1) a revised value of the strain energy difference has been found to be $\Delta E_S = 31.5$ kJ/mol (*104*) (*vs* 41.5 (*29*)).

References, pp. 56–62

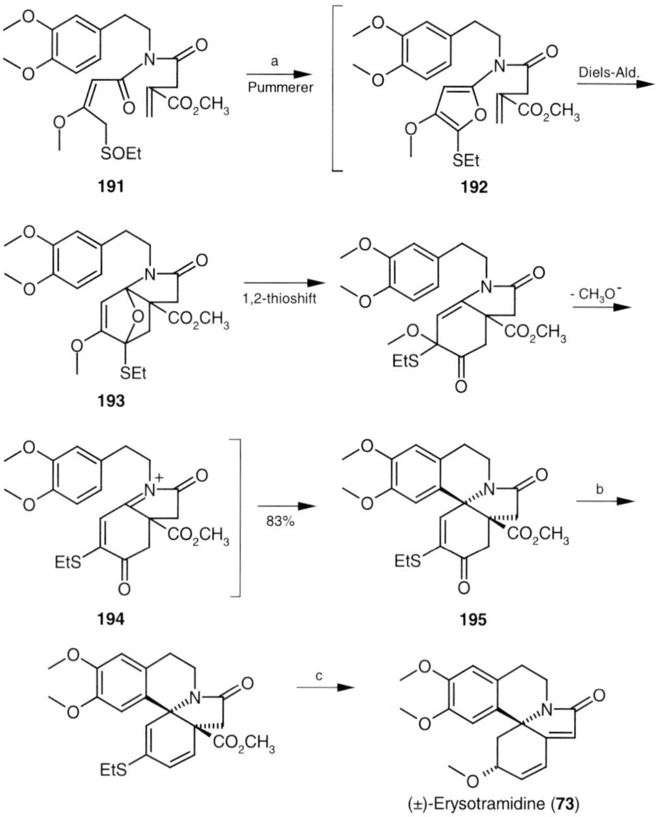

Scheme 34. Synthesis of (±)-erysotramidine (**73**) (*80*): a) Et$_3$N/(F$_3$CCO)$_2$O/Et$_2$O · BF$_3$ in CH$_2$Cl$_2$/Δ; b) *i*: KH/PhNTf$_2$ in THF; *ii*: (PPh$_3$)$_2$PdCl$_2$/NEt$_3$/HCO$_2$H in DMF/Δ; *iii*: TiCl$_4$ in AcOH; H$_2$O; c) transformation according to Ref. (*76*).

5.3.1. Biomimetic Routes

The first syntheses of the homoerythrinane framework are following the assumed biosynthetic route. Thus, the N-protected 1-phenethylisoquinoline **196** (Scheme 35) can be cyclized by a phenol oxidative procedure providing the tetracyclic naphthalenoisoquinoline **197**, a homologue of neoproaporphine (**35**) in 35–45% yield (*105*). This has been transformed to the dibenzazecine intermediate **198** by ring reopening. Finally, the N-deprotected free amine base can be oxidized giving homoerysodienone (**199**) (*106*, *107*). A diphenoquinone type intermediate **38** (Scheme 1; n = 2) is not available from the N-acylated base **198**.

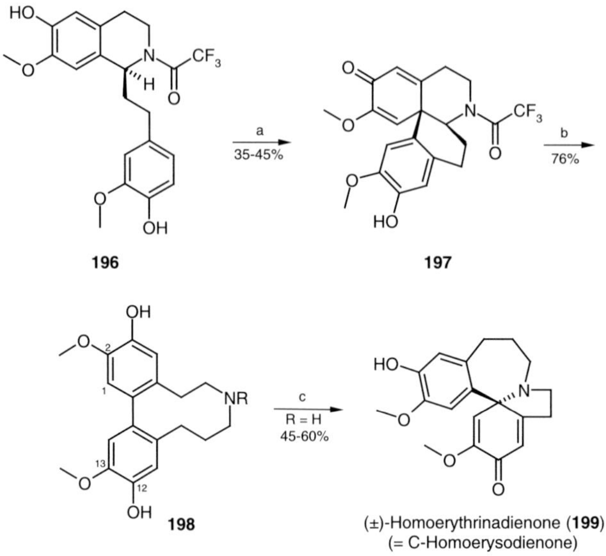

Scheme 35. Synthesis of homoerysodienone (**199**) (*105–107*): a) VOCl$_3$ in CH$_2$Cl$_2$; b) *i*: *N* NaOH in CH$_3$OH; *ii*: NaBH$_4$ in EtOH; c) K$_3$[Fe(CN)$_6$]/NaHCO$_3$ in CH$_2$Cl$_2$

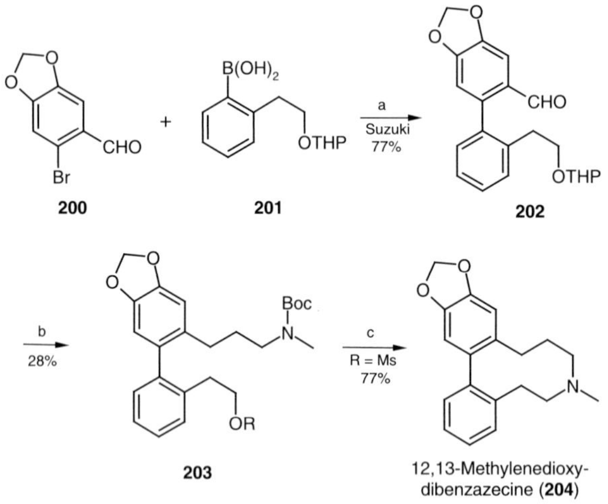

Scheme 36. Synthesis of 12,13-methylenedioxydibenzazecine (**204**) (*109*): a) Pd(OAc)$_2$/PPh$_3$/NaHCO$_3$ in dioxane/H$_2$O/reflux; b) *i*: (EtO)$_2$P(O)CH$_2$CONHCH$_3$/(C$_6$H$_{13}$)$_4$NI in NaOH; *ii*: Raney-Ni in CH$_3$OH/10^5 Pa; *iii*: LiAlH$_4$ in THF/reflux, then Boc$_2$O/Et$_3$N in CH$_2$Cl$_2$; c) *i*: AcOH in THF/H$_2$O/Δ, then MsCl/Et$_3$N in CH$_2$Cl$_2$; *ii*: F$_3$CCO$_2$H/20°C, then *i*-Pr$_2$NEt in CH$_3$CN/reflux

References, pp. 56–62

An analogous synthesis using dibenzazecine with unsymmetrically substituted aryl nuclei, *e.g.* the 3,12-dihydroxy-13-methoxy derivative (**198**, 2-H instead of 2-CH$_3$O; Scheme 35) is limited *a priori* because its inefficient preparation method leads to undesirable regioisomers (*108*). To overcome these limitations a Suzuki coupling procedure most recently developed should be particularly suitable for a more general approach to the mentioned dibenzazecine educts inclusively to their enantiomers, which offer an interesting new pathway to homoerythrinanes with flexible substitution patterns. Thus, 2-bromopiperonal (**200**) is reacted with the boronic acid **201** providing the Suzuki coupled biphenyl **202** (Scheme 36). Chain extension using the Wadsworth-Emmons method, followed by several transformation steps give the N-protected amine **203**, which can be cyclized to the target 12,13-methylenedioxydibenzazecine (**204**) (*109*).

5.3.2. Final C Ring Formation Starting from N-substituted Phenylhydroindoles

In contrast, the erythrinane synthesis based on the above mentioned cycloadditions of 1,3-butadienes to pyrrole derivatives is fully applicable to that of the ring C-homologue alkaloids. Thus, the phenylhydroindol **100b** prepared by [2 + 2] cycloaddition according to Scheme 13 cyclizes to yield the 2,8-dioxohomoerythrinane **205**, which then has been transformed via the 1,7-cyclointermediate **206** to 2,7-dihydrohomoerysotrine (**207**) (*110*) (Scheme 37).

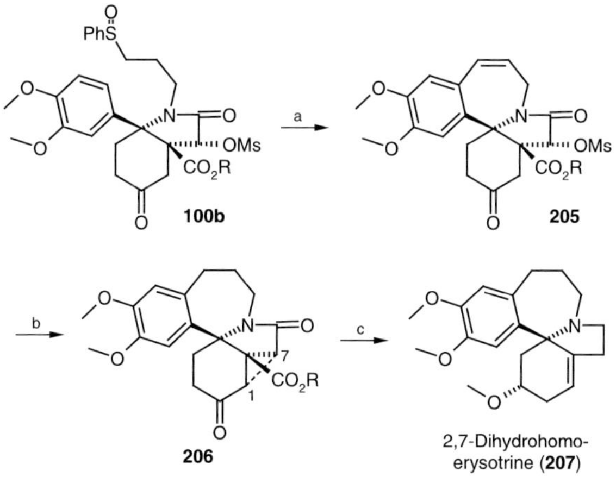

Scheme 37. General synthesis of homoerythrinanes (**207**) (*110*): a) TFAA/TFA in CH$_2$Cl$_2$/20°C; b) *i*: DBU in toluene/Δ; *ii*: H$_2$/Pd in THF; c) according to Ref. (*77*)

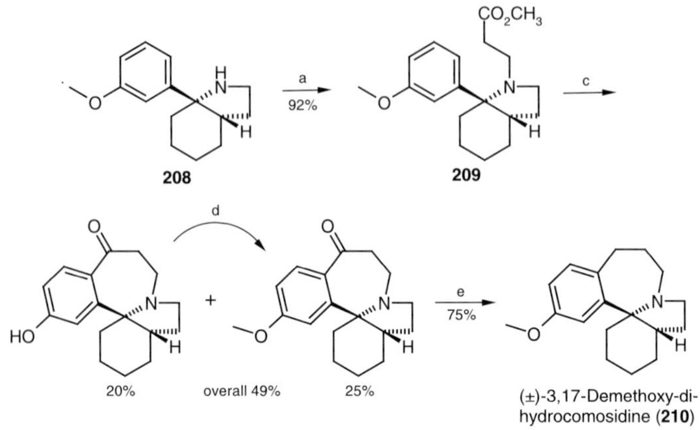

Scheme 38. Synthesis of (±)-3,17-demethoxy-1,2-dihydrocomosidine (**210**) (*28*): a) methyl acrylate in CHCl$_3$/50°C; c, d, e: reagents are the same as in Scheme 14

Furthermore, the formation of the C ring can be also accomplished following the Friedel-Crafts method already indicated in the erythrinane series (Scheme 14). Accordingly, the *cis*-arylhydroindole **208** has been alkylated with methyl acrylate providing the N-substituted precursor **209** in excellent yield, which after cyclization and consecutive removal of the carbonyl group affords the (±)-3,17-demethoxy-1,2-dihydrocomosidine **210** (Scheme 38). As already accentuated in the erythrinane series (Scheme 14) attempted approaches to the corresponding parent compound as well as to the 16,17-dioxygenated products also failed by this sequence (*28*).

5.3.3. A Ring Formation by [2 + 2] Photocycloaddition to Pyrrolobenzazepines

Similarly, replacement of the dioxopyrroloisoquinoline **137a** by the homologous dioxopyrrolobenzazepine **137b**, the synthesis of the erythrinane derivative **147a** by the [2 + 2] photocyclization method according to Scheme 21 could be successfully transferred to that of the dioxoschelhammerane **211** (Scheme 39). This in turn leads to the key compound, the α,β-unsaturated ketone **214** via the intermediates **212** and **213**, which – after generation of the 3-methoxy-derivative **215** – has been alternatively transformed to the 3-epimeric schelhammeridines **7a** and **7b** or to the corresponding 6,7-dihydro derivatives comosine (3α-**216a**) and dihydroschelhammeridine (3β-**216b**) (*78*).

The synthesis of the 3-epimeric schelhammericines, being the first synthesis approach to a homoerythrinane alkaloid at all (*111*), is based

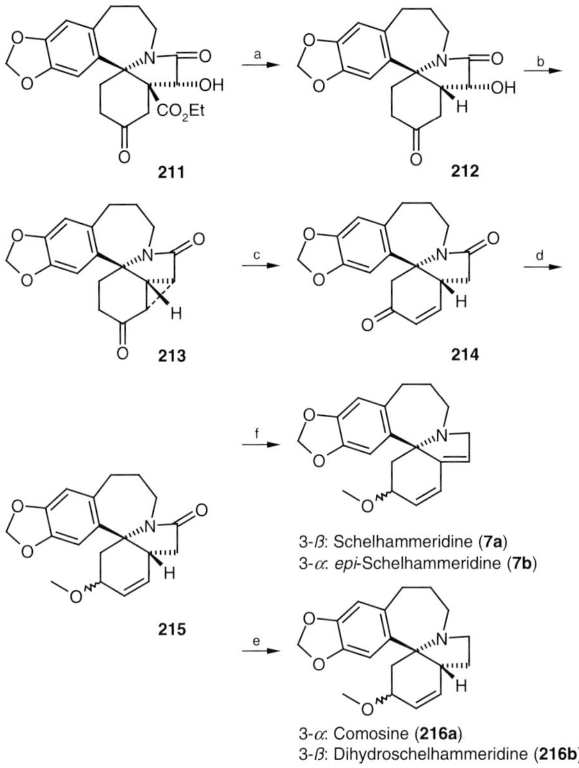

Scheme 39. Synthesis of schelhammeridines (**7**, **216**) (*78*); a) *i*: DMSO in Ac$_2$O; *ii*: ethyleneglycol/*p*-TsOH; *iii*: MgCl$_2$/Et$_3$CSH in DMSO/Δ; *iv*: NaBH$_4$, then HCl in acetone; b) MsCl in pyr, then K$_2$CO$_3$ in CH$_3$OH; c) *i*: PhSeCl/Et$_2$O · BF$_3$; *ii*: Hg(ClO$_4$)$_2$ in CH$_3$OH; *iii*: NaBH$_4$; *iv*: NaH/CS$_2$/CH$_3$I; *iw*: Bu$_3$SnH, then HCl in acetone; d) *i*: NaBH$_4$/CeCl$_3$ in CH$_3$OH, separation of 3α/3β-OH-stereomers; *ii*: NaH/Bu$_4$NHSO$_4$/CH$_3$I; e) LiAlH$_4$/AlCl$_3$, separate reduction of stereomers; f) *i*: *n*-BuLi/(PhSe)$_2$; *ii*: NaIO$_4$; *iii*: LiAlH$_4$/AlCl$_3$

on the same strategy concerning the intermediate **211** as shown in Scheme 40. This has been converted via the 1,7-cycloderivative **217** to the 2,8-dioxoschelhammerane **218**, which after consecutive stereoselective reduction, O-alkylation, and removal of lactam-oxygen affords the desired 3α- and 3β-methoxyschelhammericines **49** and **10** (*77*).

On the other hand, as already mentioned in the introduction to this section, the Diels-Alder addition of 1,3-butadienes to dioxopyrrolobenzazepine, *e.g.* according to reaction: **136** + **137b** → **138b** (Scheme 19) gives only very poor yields (<5%). This has been attributed to the marked conformational difference between the two dioxopyrroline dienophiles **137a** and **137b** (*77*).

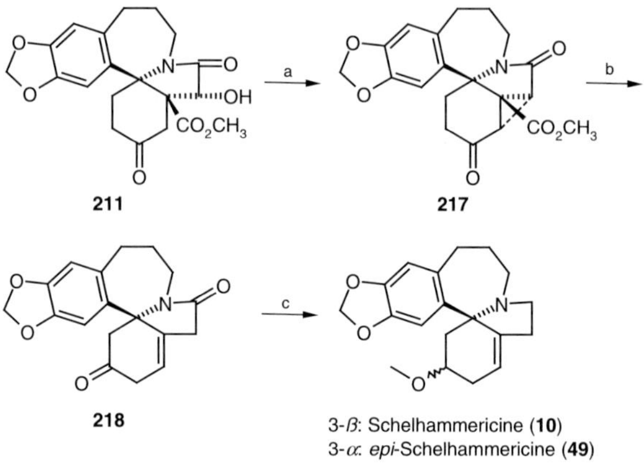

Scheme 40. Synthesis of schelhammericines (**10** and **49**) (*77, 111*): a) *i*: MsCl in pyr/20°C; *ii*: DBU in toluene/reflux; b) *i*: PhSeCl/Et$_2$O · BF$_3$ in THF/reflux, then Hg(ClO$_4$)$_2$ in CH$_3$OH; *ii*: NaBH$_4$ in CH$_3$OH/THF, 20°C; *iii*: NaH/imidazole in THF/reflux, then CH$_3$I/ CS$_2$; *iv*: Bu$_3$SnH/AIBN in toluene/reflux, then HCl in acetone; *iw*: CaCl$_2$ in DMSO/Δ; c) *i*: Bu$_4$NBH$_4$ or NaBH$_4$/CeCl$_3$ in CH$_3$OH/0°C; *ii*: NaH/imidazole/CH$_3$I/Bu$_4$NSO$_4$ in THF/reflux; *iii*: AlH$_3$ in THF/Et$_2$O/20°C

5.3.4. Simultaneous B Ring Formation/C Ring Expansion Starting from Spiro-2-tetralones

An original approach to the homoerythrinane framework represents the one pot tandem or domino alkylation-Michael addition sequence of 2-tetralones, followed by N-insertion via intramolecular

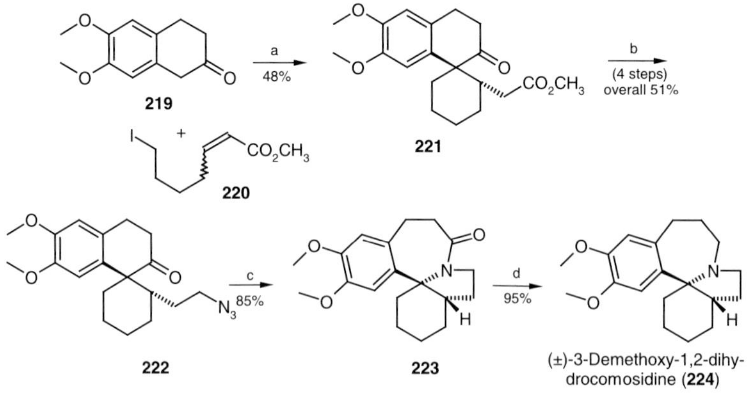

Scheme 41. Synthesis of (±)-demethoxy-1,2-dihydrocomosidine (**224**) (*112*): a) Cs$_2$CO$_3$ in DMF/20°C; b) *i*: TMSiOTf/Et$_3$N in CH$_2$Cl$_2$/20°C; *ii*: LiAlH$_4$; *iii*: MsCl; *iv*: NaN$_3$; c) neat TFA/30°C; d) LiAlH$_4$ in THF/20°C

References, pp. 56–62

Schmidt rearrangement of an azido ketone. Thus, 6,7-dimethoxy-2-tetralone (**219**) has been reacted with methyl (*E*)-7-iodo-2-heptenoate (**220**) affording stereoselectively the C1-spiroketone **221** in 48% yield. After transformation of the latter to the azido derivative **222** the tetracyclic lactam **223** was obtained under ring expansion possessing the schelhammeridine core. Finally, LiAlH$_4$ reduction smoothly provides the ($\pm$)-3-demethoxy-1,2-dihydrocomosidine base **224** (*112*) (Scheme 41).

6. Pharmacology

The biological activity of extracts of *Erythrina* species is known for a long time. Thus, concentrated extracts were used as arrow poisons by the natives of South America, as antidote against strychnine, or as hypnoticum and epilepticum (*4*). On the other hand when an alcoholic seed extract from *Erythrina americana* was applied to dogs in different doses an activity similar to that of d-tubocurarine was already observed in 1877 (*2, 113*) and confirmed much later (*7, 114–118*).

The first crystalline pharmacologically active alkaloid was isolated from *Erythrina americana* mentioned above (*7*) and was called erythroidine. This name had already been used by Altamirano referring, however, to the unknown constituents of the plant (*114*). Further analytical investigations revealed the material isolated to be a mixture of isomeric alkaloids, which were subsequently named α- and β-erythroidines (**14** and **225**, see Figs. 2 and 4).

Between 1940 and 1950 the systematic examination of more than fifty *Erythrina* species showed that all the alkaloids isolated produce effects similar to curare alkaloids (*116, 119*), which had been used as adjuvant in surgical anaesthesia (*120*).

A considerable number of properties, biological effects, and applications attributed to extracts from various *Erythrina* species has been

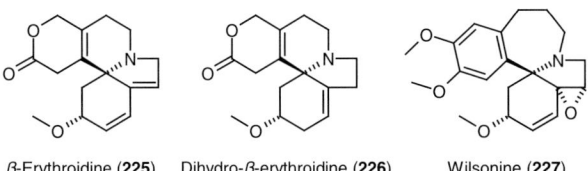

β-Erythroidine (**225**) Dihydro-β-erythroidine (**226**) Wilsonine (**227**)

Fig. 4. Chemical structures of several *Erythrina* alkaloids with special pharmacological properties (*cf.* pages 26–27)

described. Thus, they are used in indigeneous medicine as eyewashes, dressings for open wounds, for pain relieving in joints, or as calmers and relaxants (*121*). Antiasthmatic, diuretic, and hypnotic properties have also been reported. Besides their smooth curare like muscle relaxation the total alkaloids from *E. variegata* have been found CNS depressant and anticonvulsive, furthermore they increase the pentobarbital hypnosis, and inhibit the acetylcholine-induced spasm. However, these alkaloids do not have any analgesic, antipyretic, antiinflammatory, laxative, and diuretic effects. Extracts from *E. velutina*, and *E. suberosa* exhibit spasmolytic and antineoplastic activities (*19, 21, 33, 39, 122–124*). In recent investigations crude extracts of *E. americana* as well as its pure constituents β-erythroidine (**225**) and dihydro-β-erythroidine (**226**; Fig. 4) have been found to diminish the aggressive behaviour of rats using diazepam as a control. These effects are attributed to an interaction between the cholinergic and GABAergic system. Additionally, the lethal doses LD_{50} of the extracts as well as those of the pure alkaloids have been determined (*121*).

From all the pure alkaloids tested α- and β-erythroidine exhibit the highest activity (*121*). They have been assumed to be the principles responsible for the hypnotic activity of the extracts of the flowers of *E. americana* (*19*). β-Eyrthroidine (**225**) and its more potent 2,7-dihydro-derivative **226** have been used as muscle relaxant in numerous clinical applications (*114, 120*). This activity is attributed to an antagonistic action of dihydro-β-erythroidine to nicotinic acetylcholine receptors, which is now well known and is frequently used in the experimental pharmacology (*19, 125, 126*).

Erysodine (**6**) has been found to be a competitive, reversible antagonist of nicotine-induced dopamine release. It is equipotent with dihydro-β-erythroidine (**226**) and may be a useful tool to characterize neuronal nicotinic acetylcholine receptors (*127*).

Very little is known about the pharmacological action of homoerythrinanes. Indeed, they are occurring together with *Cephalotaxus* alkaloids and are biosynthetically and structurally closely related to them, but nevertheless, they seem to possess no antitumor properties like the latter. Wilsonine (**227**; Fig. 4), however, shows weak antileukemic effects in mice (*128*).

3-*Epi*-schelhammericine (**49**), and dyshomoerythrine (**10**: 18-OCH$_3$ instead of 18-H) are known to exhibit a strong molluscicidal activity; the latter is also active against some agriculturally important insect pests (*129*). Lenticellarine (**32**: 2-H instead of 2-OH) occurring in the same plant is less effective. 3-*Epi*-schelhammericine and dyshomoerythrine as

References, pp. 56–62

well as 2,7-dihydro-homoerysotrine (**207**) and 3-*epi*-12-hydroxyschelhammericine (**49**: 12-OH instead of 12-H) are considered to exhibit cardiovascular effects (*19*).

Finally, it should be mentioned that the curare alkaloids – as is generally known – are quaternary salts and due to their strong hydrophilicity they have to be administered parenterally. In contrast, the *Erythrina* alkaloids are all tertiary amines and therefore they are able to develop their pharmacological activities upon oral administration.

7. Concluding Remarks

The present review summarizes the work of the last fifty years in the field of the synthesis of *Erythrina* alkaloids and structurally related compounds. Their unique structure has attracted the attention of the synthesis chemists over a period of more than half a century until now. The numerous efforts devoted to the construction of the tetracyclic spiroamine framework have certainly led to original and impressive results, but with regard to today's standard and especially to an adequate supply of the promising compounds for a systematic pharmacological examination the total efficiency of their preparation yet leaves wishes in many cases.

While a great number of synthesis pathways providing the aromatic erythrinane type compounds is reported, comparatively much less approaches to the homoerythrinanes have been developed. This may be explained by the fact, that the erythrinane routes cannot be *a priori* applied to the synthesis of the corresponding C-homologues. Furthermore, asymmetric syntheses of the alkaloids are rare, and approaches to the nonaromatic and heteroaromatic compounds are lacking. There is certainly a need for shorter routes to the target compounds. For instance, domino reactions obviously represent the tool of choice for this purpose as it is demonstrated herein enabling the rapid construction of complex frameworks.

Not only because of their attractive structure, but also of their pharmacological potential, mainly that of the pure compounds, the synthesis of the *Erythrina* alkaloids will remain an important goal challenging the synthesis chemists in the forthcoming years. It is hoped, that this review along with the powerful arsenal of modern synthesis methods can stimulate the development of new efficient strategies to the *Erythrina* alkaloids and possibly to related new active agents.

References

1. Rey P (1883) Note sur les propriétés thérapeutiques de l'Érythrina corallodendron. Le Journal de Therapeutique 10: 843
2. Altamirano F (1888) Nuevos Apuntes para el Estudio del Colorin, *Erytrina coralloides*. Gaceta Médica de México 23: 369
3. Greshoff M (1890) Mitteilungen aus dem chemisch-pharmakologischen Laboratorium des Botanischen Gartens zu Buitenzorg (Java). Chem Ber 23: 3537
4. Plugge PC (1894) Untersuchung einiger niederländisch-ostindischen Pflanzenstoffe. V. Einiges über die Wirkung des Alkaloids von *Erythrina* (Stenotropis) Broteroi Hssk. Arch Exp Pathol Pharmakol 33: 46
5. Folkers K, Unna K (1938) Erythrina Alkaloids. II. A Review, and New Data on the Alkaloids of Species of the Genus Erythrina. J Am Pharm Assoc 27: 693
6. Folkers K, Unna K (1939) Erythrina Alkaloids. V. Comparative Curare-like Potencies of Species of the Genus Erythrina. J Am Pharm Assoc 28: 1019
7. Folkers K, Major RT (1937) Isolation of Erythroidine, an Alkaloid of Curare Action, from Erythrina americana Mill. J Am Chem Soc 59: 1580
8. Folkers K, Koniuszy F (1939) Erythrina Alkaloids. III. Isolation and Characterization of a New Alkaloid, Erythramine. J Am Chem Soc 61: 1232
9. Folkers K, Koniuszy F (1940) Erythrina alkaloids. VIII. Studies on the Constitution of Erythramine and Erythraline. J Am Chem Soc 62: 1673
10. Folkers K, Koniuszy F (1940) Erythrina Alkaloids. VII. Isolation and Characterization of the New Alkaloids, Erythraline and Erythratine. J Am Chem Soc 62: 436
11. Folkers K, Koniuszy F (1940) Erythrina Alkaloids. IX. Isolation and Characterization of Erysodine, Erysopine, Erysocine, and Erysovine. J Am Chem Soc 62: 1677
12. Carmack M, MacKusick BC, Prelog V (1951) Erythrina-Alkaloide, 2. Mitt. Über das Apo-erysodin und das Apo-erythralin. Helv Chim Acta 34: 1601
13. Kenner GW, Khorana HG, Prelog V (1951) Erythrina-Alkaloide, 3. Mitt. Über den *Hofmann*'schen Abbau des Tetrahydro-erysotrins und des Tetrahydro-erythralins. Helv Chim Acta 34: 1969
14. Boekelheide V, Weinstock J, Grunden MF, Sauvage GL, Agnello EJ (1953) The Structure of β-Erythroidine and Its Derivatives. J Am Chem Soc 75: 2550
15. Kametani T, Fukumoto K (1968) Synthesis of "Homoerythrinadienone" by Phenolic Oxidative Coupling. J Chem Soc Chem Commun 26
16. Kametani T, Fukumoto K (1968) Synthesis of Homoerythrinadienone by Phenolic Oxydative Coupling. J Chem Soc (C) 2156
17. Johns SR, Kowala C, Lamberton JA, Sioumis AA, Wunderlich JA (1968) "Homoerythrina" Alkaloids from *Schelhammera pedunculata* F.Muell. (Family Liliaceae). J Chem Soc Chem Commun 1102
18. Chawla AS, Kapoor VK (1997) *Erythrina* Alkaloids. In: D'Mello JPF (ed) Handbook of Plant and Fungal Toxicants. CRC Press (Erythrinane alkaloids exclusively)
19. Tsuda Y, Sano T (1996) Erythrina and Related Alkaloids. In: Cordell GA (ed) The Alkaloids vol. 48. Academic Press, Inc. (Erythrinane and Homoerythrinane alkaloids)
20. Tsuda Y, Sano T (1989) Synthesis of Erythrina Alkaloids. In: Atta-ur-Rahman (ed) Studies in Natural Products Chemistry Vol. 3, Stereoselective Synthesis (Part B) Elsevier, Amsterdam, p 455 (Erythrinane and Homoerythrinane alkaloids, approaches to non-aromatic Erythrina alkaloids)
21. Chawla AS, Kapoor VK (1995) *Erythrina* Alkaloids. In: Pelletier SW (ed) Alkaloids: Chemical and Biological Perspectives vol. 9. Pergamon (Erythrinane alkaloids exclusively)

22. Bick IRC, Panichanun S (1991) Homoerythrina and Related Alkaloids. In: Pelletier SW (ed.) Alkaloids: Chemical and Biological Perspectives Vol. 7. Springer, Berlin (Homoerythrinane alkaloids exclusively)
23. Hegnauer R, Hegnauer M (2001) Chemotaxonomie der Pflanzen Band XIb-2, Leguminosae, Teil 3: Papilionoideae. Birkhäuser, Basel, p 319 ff (Erythrininae)
24. Amer ME, Shamma M, Freyer AJ (1991) The Tetracyclic *Erythrina* Alkaloids. J Nat Prod 54: 329 (A compilation of ca. 90 erythrina-type alkaloids together with their spectral data)
25. Dagne E, Steglich W (1983) Erymelanthine, a New Type of Erythrina Alkaloid Containing a 16-Azaerythrinane Skeleton. Tetrahedron Lett 24: 5067
26. Langlois N, Razafimbelo J, Andriamialisoa RZ, Pusset J, Chauviere G (1984) Alkaloids Isolated from the Leaves of *Phelline* sp. aff. *P. lucida* (Phellinaceae). Heterocycles 22: 2453
27. Mondon A, Hansen KF, Boehme K, Faro HP, Nestler HJ, Vilhuber HG, Böttcher K (1970) Synthetische Arbeiten in der Reihe der aromatischen Erythrina-Alkaloide, XI: Anwendungen der Glyoxylester-Synthese. Chem Ber 103: 615
28. Reimann E, Ettmayr C (2004) A Novel Stereoselective Synthesis of *cis*-Configured Erythrinane and Erythrinane Type Analogues. Monatsh Chem 135: 959
29. Reimann E, Ettmayr C (2004) An Improved Stereocontrolled Route to *cis*-Erythrinanes by Combined Intramolecular Strecker and Bruylants Reaction. Monatsh Chem 135: 1143
30. Boekelheide V, Wenzinger GR (1964) The Absolute Configurations of the Erythrina Alkaloids. J Org Chem 29: 1307
31. Mondon A, Seidel PR (1971) Synthetische Arbeiten in der Reihe der aromatischen Erythrina-Alkaloide, XX: Synthese des 15,16-Dimethoxy-*trans*-erythrinans und dessen Abbau zum *trans*-Erythrinan. Chem Ber 104: 2937
32. Jousse C, Desmaële D (1999) A New Approach to Erythrinanes through Pummerer-Type Cyclization. Eur J Org Chem 909
33. Sharma SK, Chawla HM (1998) Structure Elucidation of Erythrosotidienone and Erythromotidienone – Two New Isoquinoline Alkaloids from *Erythrina variegata* Flowers. J Indian Chem Soc 75: 833
34. Wanjala CCW, Juma BF, Bojase G, Gashe BA, Majinda RRT (2002) Erythraline Alkaloids and Antimicrobial Flavonoids from *Erythrina latissima*. Planta Med 68: 640
35. Hussain SS (2002) A New Alkaloid from Flowers of *Erythrina stricta*. J of Sciences/ Islam Rep of Iran 13: 35
36. Tanaka H, Tanaka T, Etoh H, Goto S, Terada Y (1999) Two New Erythrinan Alkaloids from *Erythrina* x *bidwillii*. Heterocycles 51: 2759
37. Tanaka H, Tanaka T, Etoh H (1998) Erythrinan Alkaloid from *Erythrina* x *bidwillii*. Phytochemistry 48: 1461
38. Amer ME, Kassem FF, El-Masry S, Shamma M, Freyer AJ (1993) NMR Spectral Analysis of Five Alkaloids from *Erythrina caffra*. Alexandria J Pharm Sci 7: 28
39. Wanjala CW, Majinda RT (2000) Two Novel Glucodienoid Alkaloids from *Erythrina latissima* Seeds. J Nat Prod 63: 871
40. Tsakadze D, Sturua M, Kupatashvilli N, Vepkhvadze T, Ziaiev R, Samsonia Sh, Abdusamatov A (1997) Alkaloids of *Cocculus laurifolius* D.C. Bull Georgian Acad Sci 155: 372
41. Tanaka H, Etoh H, Shimizu H, Oh-Uchi T, Terada Y, Tateishi Y (2001) Erythrinan Alkaloids and Isoflavonoids from *Erythrina poeppigiana*. Planta Med 67: 871

42. Langlois N, Hamon J (2004) Alcaloïdes Homoérythrina isolés de *Phelline comosa* var. *robusta*. C R Chimie 7(1): 51
43. Qiu M, Lu B, Ma X, Nei R (1997) Alkaloids from *Cephalotaxus fortunei* Collected in Lijiang. Yunnan Zhiwu-Yanjiu (= Acta Botanica Yunnanica) 19(1): 97 (Chem Abstr (1997) 127: 275 355)
44. Morita H, Yoshinaga M, Kobayashi J (2002) Cephalezomines G, H, J, K, L, and M, New Alkaloids from *Cephalotaxus harringtonia* var. *nana*. Tetrahedron 58: 5489
45. Aladesanmi AJ, Hoffmann JJ (1994) Additional Alkaloids from the Stem of *Dysoxylum lenticellare*. Phytochemistry 35: 1361
46. Barton DHR, Cohen T (1957) Some Biogenetic Aspects of Phenol Oxidation. In Festschrift Professor Dr. Artur Stoll. Basel: Birkhäuser
47. Barton DHR, Boar RB, Widdowson DA (1970) Phenol Oxidation and Biosynthesis. Part XXI. The Biosynthesis of the *Erythrina* Alkaloids. J Chem Soc C 1213
48. Barton DHR, Potter CJ, Widdowson DA (1974) Phenol Oxidation and Biosynthesis. Part XXIII. On the Benzyltetrahydroisoquinoline Origins of the *Erythrina* Alkaloids. J Chem Soc Perkin Trans 1 346
49. Ghosal S, Majumdar SK, Chakraborti A (1971) *Erythrina* Alkaloids, III. Occurrence of (+)-*N*-Norprotosinomenine and Other Alkaloids in *Erythrina lithosperma* (Leguminosae). Aust J Chem 24: 2733
50. Aladesanmi AJ, Kelley CJ, Leary JD (1983) The Constituents of *Dysoxylum lenticellare*, I. Phenylethylisoquinoline, Homoerythrina, and Dibenzazecine Alkaloids. J Nat Prod 46: 127
51. Franck B, Teetz V (1971) Modellreaktionen zur Biosynthese der Erythrina-Alkaloide. Angew Chem 83: 409
52. Maier HU, Rödl W, Deus-Neumann B, Zenk MH (1999) Biosynthesis of *Erythrina* Alkaloids in *Erythrina crista-galli*. Phytochemistry 52: 373
53. Barton DHR, James R, Kirby GW, Turner DW, Widdowson DA (1968) Phenol Oxidation and Biosynthesis. Part XVIII. The Structure and Biosynthesis of *Erythrina* Alkaloids. J Chem Soc C 1529
54. Kawasaki T, Onoda N, Watanabe H, Kitahara T (2001) Total Synthesis of ($\pm$)-Cocculolidine. Tetrahedron Lett 42: 8003
55. Dyke SF, Quessy SN (1981) *Erythrina* and Related Alkaloids. In: Rodrigo RGA (ed) The Alkaloids Vol. 18. Academic Press, Inc., New York
56. Mondon A (1958) Neue Synthesen auf dem Gebiet der aromatischen Erythrina-Alkaloide. Angew Chem 70: 406
57. Mondon A (1959) Synthetische Arbeiten in der Reihe der aromatischen Erythrina-Alkaloide, IV: Synthese ungesättigter Erythrinane. Liebigs Ann Chem 628: 123
58. Tsuda Y, Sakai Y, Kaneko M, Ishiguro Y, Isobe K, Taga J, Sano T (1981) A Practical Route to Spiro-Type Heterocycles Related to Erythrinan. Heterocycles 15: 431
59. Tsuda Y, Sakai Y, Nakai A, Kaneko M, Ishiguro Y, Isobe K, Taga J, Sano T (1990) Synthesis of Erythrina and Related Alkaloids. XXII. Intramolecular Cyclization Approach. (1): New Synthetic Route to Erythrinan and Related Heterocycles and Synthesis of ($\pm$)-3-Demethoxyerythratidinone. Chem Pharm Bull 38: 1462
60. Tsuda Y, Hosoi S, Ishida K, Sangai M (1994) Chiral Synthesis of *Erythrina* Alkaloids (2). Synthesis of *enantio*-Type Erythrinan Alkaloids Utilizing Asymmetric Acylation and Kinetic Resolution of Diastereomers. Chem Pharm Bull 42: 204
61. Allin SM, James SL, Elsegood MRJ, Martin WP (2002) Facile and Highly Stereoselective Synthesis of the Tetracyclic Erythrinane Core. J Org Chem 67: 9464

62. a) Lee HI, Cassidy MP, Rashatasakhon P, Padwa A (2003) Efficient Synthesis of (±)-Erysotramidine Using an NBS-Promoted Cyclization Reaction of a Hexahydroindolinone Derivative. Org Lett 5: 5067; b) Padwa A, Lee HI, Rashatasakhon P, Rose M (2004) Electrophilic-Induced Cyclization Reaction of Hexahydroindolinone Derivatives and Its Application Toward the Synthesis of (±)-Erysotramidine. J Org Chem 69: 8209
63. Cassayre J, Quiclet-Sire B, Saunier JB, Zard SZ (1998) Nickel Powder Promoted 5-*endo* Radical Cyclisations. A Concise Approach to *Erythrina* Alkaloids. Tetrahedron Lett 39: 8995
64. Rigby JH, Deur C, Heeg MJ (1999) Synthetic Studies on the Erythrina Alkaloids. Preparation of (±)-2-*epi*-Erythrinitol. Tetrahedron Lett 40: 6887
65. Toda J, Niimura Y, Takeda K, Sano T, Tsuda Y (1998) General Method for Synthesis of Erythrinan and Homoerythrinan Alkaloids (1). Synthesis of a Cycloerythrinan, as a Key Intermediate to *Erythrina* Alkaloids, by Pummerer-Type Reaction. Chem Pharm Bull 46: 906
66. Zhang Y, Takeda S, Kitagawa T, Irie H (1986) A Synthesis of 15,16-Dimethoxyerythrin-6-en-8-one. Heterocycles 24: 2151
67. Irie H, Shibata K, Matsuno K, Zhang Y (1989) Synthesis of an Alkaloid, (±)-3-Demethoxyerythratidinone. Heterocycles 29: 1033
68. Giese B, Horler H, Zwick W (1982) Synthesis of 1,6-Heterosubstituted Products via Radicals. Tetrahedon Lett 23: 931
69. Danishefsky SJ, Panek JS (1987) Total Synthesis of (±)-3-Demethoxyerythratidinone: Demonstration of a Radical Cyclization Route to a Site Specific Enol Derivative. J Am Chem Soc 109: 917
70. Ahmed-Schofield R, Mariano PS (1987) A Photochemical Route for Erythrane Ring Construction. J Org Chem 52: 1478
71. Chou CT, Swenton JS (1987) A Convergent Strategy for Synthesis of *Erythrina* Alkaloids. J Am Chem Soc 109: 6898
72. Sano T, Toda J, Kashiwaba N, Ohshima T, Tsuda Y (1987) Synthesis of *Erythrina* and Related Alkaloids, XVI. Diels-Alder Approach: Total Synthesis of *dl*-Erysotrine, *dl*-Erythraline, *dl*-Erysotramidine, *dl*-8-Oxoerythraline and Their 3-Epimers. Chem Pharm Bull 35: 479
73. Tsuda Y, Hosoi S, Katagiri N, Kaneko C, Sano T (1992) Total Synthesis of (+)-Erysotrine *via* Asymmetric Diels-Alder Reaction Under Super High Pressure. Heterocycles 33: 497
74. Tsuda Y, Hosoi S, Katagiri N, Kaneko C, Sano T (1993) Chiral Synthesis of *Erythrina* Alkaloids. I: Total Synthesis of (+)-Erysotrine *via* Asymmetric Diels-Alder Reaction Under High Pressure. Chem Pharm Bull 41: 2087
75. Sano T, Toda J, Ohshima T, Tsuda Y (1992) Synthesis of *Erythrina* and Related Alkaloids. XXX. Photochemical Approach. (1). Synthesis of Key Intermediates to *Erythrina* Alkaloids by Intermolecular [2+2] Photocycloaddition Followed by 1,3-Shift. Chem Pharm Bull 40: 873
76. Tsuda Y, Hosoi S, Nakai A, Sakai Y, Abe T, Ishi Y, Kiuchi F, Sano T (1991) Synthesis of *Erythrina* and Related Alkaloids. XXIV. Total Synthesis of Erysotrine from 1,7-Cycloerythrinan Derivatives by the Use of a New 1,2-Carbonyl Transposition Method. Chem Pharm Bull 39: 1365
77. Tsuda Y, Ohshima T, Hosoi S, Kaneuchi S, Kiuchi F, Toda J, Sano T (1996) Total Synthesis of Homoerythrinan Alkaloids, Schelhammericine and 3-Epischelhammericine. Chem Pharm Bull 44: 500
78. Tsuda Y, Murata M, Hosoi S, Ikeda M, Sano T (1996) Synthesis of Homoerythrinan Alkaloids of 1(2)-Alkene and 1,6-Diene Types: Total Synthesis of Comosine,

Dihydroschelhammeridine, Schelhammeridine, and 3-Epischelhammeridine. Chem Pharm Bull 44: 515
79. Wasserman HH, Amici RM (1989) The Chemistry of Vicinal Tricarbonyls. A Total Synthesis of ($\pm$)-3-Demethoxyerythratidinone. J Org Chem 54: 5843
80. Padwa A, Hennig R, Kappe CO, Reger TS (1998) A Triple Cascade Sequence as a Strategy for the Construction of the Erythrinane Skeleton. J Org Chem 63: 1144, and lit cited therein
81. Belleau B (1953) The Synthesis of Erythrinane. J Am Chem Soc 75: 5765
82. Belleau B (1957) Synthesis in the Field of the Erythrina Alkaloids. Part I. The Synthesis of Hexahydroapoerysotrine. Can J Chem 35: 651
83. Mondon A (1956) Synthese des 15,16-Dimethoxy-erythrinans. Angew Chem 68: 578
84. Mondon A, Hasselmeyer G, Zander J (1959) Synthetische Arbeiten in der Reihe der aromatischen Erythrina-Alkaloide, III. Studien zum Ringschluß in der Erythrinan-Reihe. Chem Ber 92: 2543
85. Mondon A (1959) Synthetische Arbeiten in der Reihe der aromatischen Erythrina-Alkaloide, I. Synthese des *racem. trans*-15,16-Dimethoxy-erythrinans. Chem Ber 92: 1461
86. Mondon A, Böttcher K (1970) Synthetische Arbeiten in der Reihe der aromatischen Erythrina-Alkaloide, XIII. Ein direkter Weg zum 15,16-Dimethoxy-*cis*-erythrinandion-(1.8). Chem Ber 103: 1512
87. Mondon A, Ehrhardt M (1966) Eine Totalsynthese des Dihydroerysodins nach biogenetischem Vorbild. Tetrahedron Lett 23: 2557
88. Gervay JE, McCapra F, Money T, Sharma GM (1966) Phenol Oxidation. A Model for the Biosynthesis of the *Erythrina* Alkaloids. J Chem Soc Chem Commun 142
89. Tanaka H, Shibata M, Ito K (1984) Synthesis of 3-Demethoxyerythratidinone *via* Formation of a Dibenzazonine Alkaloid. Chem Pharm Bull 32: 1578
90. Haruna M, Ito K (1976) Convenient Method for Syntheses of Erythrinan Alkaloids. J Chem Soc Chem Commmun 345
91. Tanaka H, Shibata M, Ito K (1984) A Novel Synthesis of *cis*-15,16-Dimethoxyerythrinan-3-one. Chem Pharm Bull 32: 3271
92. El Bialy SAA, Braun H, Tietze LF (2004) A Highly Efficient Synthesis of the Erythrina and B-Homoerythrina Skeleton by an AlMe$_3$-Mediated Domino Reaction. Angew Chem Int Ed 43: 5391
93. Padwa A, Gunn DE, Osterhout MH (1997) Application of the Pummerer Reaction Toward the Synthesis of Complex Carbocycles and Heterocycles. Synthesis 1997, 1353
94. Ishibashi H, Sato K, Ikeda M, Maeda H, Akai S, Tamura Y (1985) One-step Synthesis of the Erythrinane Skeleton by Acid-promoted Double Cyclization of *N*-(Cyclohex-1-enyl)-*N*-[2-(3,4-dimethoxyphenyl)ethyl]-α-(methylsulphinyl)acetamide and Its Derivatives. J Chem Soc Perkin Trans 1 605
95. Padwa A, Waterson AG (2000) Studies Dealing with Thionium Ion Promoted Mannich Cyclization Reaction. J Org Chem 65: 235
96. Miranda LD, Zard SZ (2002) A Short Synthesis of the Erythrina Skeleton and of ($\pm$)-α-Lycorane. Org Lett 7: 1135
97. Chikaoka S, Toyao A, Ogasawara M, Tamura O, Ishibashi H (2003) Mn(III)/Cu(II)-Mediated Oxidative Radical Cyclization of α-(Methylthio)acetamides Leading to Erythrinanes. J Org Chem 68: 312
98. Ishibashi H, Sato T, Takahashi M, Hayashi M, Ishikawa K, Ikeda M (1990) A Concise Total Synthesis of ($\pm$)-3-Demethoxyerythratidinone Based on an Acid-Promoted Double Cyclization of α-Sulfinylacetamides. Chem Pharm Bull 38: 907

99. Westling M, Smith R, Livinghouse T (1986) A Convergent Approach to Heterocycle Synthesis via Silver Ion Mediated α-Keto Imidoyl Halide-Arene Cyclizations. An Application to the Synthesis of the Erythrinane Skeleton. J Org Chem 51: 1159
100. Ishibashi H, Harada S, Sato K, Ikeda M, Akai S, Tamura Y (1985) Synthesis of the Erythrinan Skeleton by Acid-Promoted Cyclization of N-(3-Oxo-1-cyclohexen-1-yl)-N-[2-(3,4-dimethoxyphenyl)ethyl]-α-(methylsulfinyl)acetamide. Chem Pharm Bull 33: 5278
101. Tsuda Y, Oshima T (1982) Δ^2-Pyrroline-4,5-dione, an Ambient Dienophile in Diels-Alder Reaction. Heterocycles 19: 2053
102. PCMODEL V 7.5, Molecular Modeling Software: Serena Software, Box 3076, Bloomington, IN 47402-3076, USA
103. Eliel EL, Wilen SH (1994) Stereochemistry of Organic Compounds. Wiley, New York, p 764
104. Reimann E, Ettmayr C: unpublished results
105. Marino JP, Samanen JM (1973) The Chemistry of Prohomoerythrinadienone I. Tetrahedron Lett 4553
106. Marino JP, Samanen JM (1976) A Biogenetic-Type Approach to Homoerythrina Alkaloids. J Org Chem 41: 179
107. McDonald E, Suksamrarn A (1975) Total Synthesis of Compounds Related to the Homoerythrina Alkaloids. Tetrahedron Lett 4425
108. Tanaka H, Tamura Y, Shibata M, Ito K (1986) A Novel Conversion of Dibenzazecines into Homoerythrina Compounds: Synthesis of cis-16,17-Dimethoxyhomoerythrinan-3-one. Chem Pharm Bull 34: 24
109. Furneaux RH, Gainsford GJ, Mason JM (2004) Synthesis of the Enantiomers of Hexahydrodibenz[d,f]azecines. J Org Chem 69: 7665
110. Toda J, Niimura Y, Sano T, Tsuda Y (1998) General Method for Synthesis of Erythrinan and Homoerythrinan Alkaloids (2): Application of Pummerer-Type Reaction to the Synthesis of Homoerythrinan Ring System. Heterocycles 48: 1599
111. Tsuda Y, Hosoi S, Oshima T, Kaneuchi S, Murata M, Kiuchi F, Toda J, Sano T (1985) Total Synthesis of the Homoerythrinan Alkaloids, Schelhammericine and 3-Epischelhammericine. Chem Pharm Bull 33: 3574
112. Le Dréau MA, Desmaële D, Dumas F, d'Angelo J (1993) A New Access to Homoerythrina Alkaloids. J Org Chem 58: 2933
113. Dominguez M, Altamirano F (1877) Del Colorin. Gaceta Médica de México 12: 77
114. Lozoya X, Lozoya M (1982) Flora medicinal de México. In Plantas Indígenas del Seguro Social México, DF p 174
115. Lehman AJ (1936) Curare-Actions of Erythrina americana. Proc Soc Exp Biol Med 33: 501
116. Lehman AJ (1937) Action of Erythrina americana, a Possible Curare Substitute. J Pharmacol Exp Ther 60: 69
117. Ramirez E, Rivero MD (1935) Pharmacodynamic Action of Erythrina americana Mill [family Leguminoseae]. Anales Inst. Biol. (Mex.) 6: 301 (Chem Abstr (1936) 30: 3088[9])
118. Garcia-Mateos R, Garin-Aguilar ME, Soto-Hernandez M, Martinez-Vasquez M (2000) Effect of β-Erythroidine and β-Dihydroerythroidine from Erythrina americana on Rats Aggressive Behaviour. Pharm Pharmacol Lett 10: 34
119. Pick EP, Unna K (1945) The Effect of Curare and Curare-like Substances on the Central Nervous System. J Pharmacol Exp Therap 83: 59
120. Payne LG (1991) The Alkaloids of Erythrina. Clonal Evaluation and Metabolic Fate. PhD Thesis. Department of Chemistry, Louisana State University, USA

121. Garin-Aguilar ME, Ramirez Luna JE, Soto-Hernandez M, del Toro GV, Vazquez MM (2000) Effect of Crude Extracts of *Erythrina americana* Mill. on Aggressive Behavior in Rats. J Ethnopharmacol 69: 189
122. Ghosal S, Dutta S, Bhattacharya SK (1972) Erythrina – Chemical and Pharmacological Evaluation. II: Alkaloids of *Erythrina variegata* l. J Pharm Sci 61: 1274
123. Hargreaves RT, Johnson RD, Millington DS, Mondal MH, Beavers W, Becker L, Young C, Rinehart KL jr (1974) Alkaloids of American Species of *Erythrina*. Lloydia 37: 569
124. Craig LE (1955) Curare-like Effects. In: Manske RHF (ed) The Alkaloids; Chemistry and Physiology. Vol. 5 – Pharmacology. Academic Press, New York, p 265 ff
125. Cheeta S, Tucci S, File SE (2001) Antagonism of the Anxiolytic Effect of Nicotine in the Dorsal Raphé Nucleus by Di-hydro-β-erythroidine. Pharmacol Biochem Behavior 70: 491
126. Schoffelmeer ANM, De Vries TJ, Wardeh G, van de Ven HWM, Vanderschuren LJMJ (2002) Psychostimulant-Induced Behavioral Sensitization Depends on Nicotinic Receptor Activation. J Neurosci 22: 3269
127. Decker MW, Anderson DJ, Brioni JD, Donnelly-Roberts DL, Kang CH, O'Neill AB, Piattoni-Kaplan M, Swanson S, Sullivan JP (1995) Erysodine, a Competitive Antagonist at Neuronal Nicotinic Acetylcholine Receptors. Eur J Pharmacol 280: 79
128. No authors given (1980) Studies on Alkaloids of *Cephalotaxus sinensis* (Rehd. et Wils.) Li. Zhiwu Xuebao 22: 156 (Chem Abstr (1980) 93: 235 142)
129. Hart JB, Mason JM, Gerard PJ (2001) Semi-synthesis and Insecticidal Activity of Dyshomoerythrine Derivatives. Tetrahedron 57: 10033

The Trichothecenes and Their Biosynthesis

J. F. Grove[†]

3 Homestead Court, Welwyn Garden City, Herts AL7 4LY, England

Contents

1. Introduction 63
2. The Trichothecenes 64
   2.1. Macrocyclic and Non-Macrocyclic Compounds 64
   2.2. Trichothecene Relatives 90
   2.3. Sources. 96
   2.4. Oxygenation Pattern 97
3. Biosynthesis. 98
   3.1. Simple Trichothecenes 98
      3.1.1. Mevalonic Acid to Trichodiene 98
      3.1.2. Trichodiene to 12,13-Epoxytrichothecene and Isotrichodermol 101
      3.1.3. Further Oxygenation and Esterification of the Trichothecene
           Nucleus: Biosynthesis of Specific Metabolites 104
           3.1.3.1. Trichothecolone 104
           3.1.3.2. Vomitoxin and Derivatives 104
           3.1.3.3. T-2 Toxin 107
           3.1.3.4. Nivalenol and Derivatives 108
      3.1.4. Trichothecene Biosynthetic Gene Clusters 108
   3.2. Trichoverroids and Macrocyclic Trichothecenes 109
   3.3. Trichothecene Relatives 112

References 113

1. Introduction

The trichothecenes are a group of naturally-occurring sesquiterpenoid epoxides which show a broad range of biological activity. They are powerful inhibitors of eukaryotic protein synthesis, are phytotoxic, insecticidal and toxic to animals, and some are among the most toxic non-nitrogenous compounds known to man. Several are commonly found in cereal grains, and the potential health risk from contaminated animal feed and human food is a major factor in stimulating research into this

---
[†]Deceased.

group of compounds. Since the isolation of the macrolide mixture, glutinosin (*1*), was reported in 1946, the body of trichothecene literature has expanded to some 3000 publications.

This review lists the trichothecenes recorded in the literature up to Dec. 2000, together with their sources, coupled with a summary of the pathways involved in trichothecene biosynthesis, an area in which significant advances have been made during the past decade. The review contains both macrocyclic trichothecenes ("macrocycles"), previously recorded to Dec. 1991 (*1*), and non-macrocyclic trichothecenes, previously recorded to Dec. 1995 (*2, 3*). Some omissions from these earlier lists have been included, and some errors corrected.

2. The Trichothecenes

2.1. Macrocyclic and Non-Macrocyclic Compounds

A total of 217 trichothecenes, based on the sesquiterpene skeleton (**1**) named trichothecane (*4*), which replaced the earlier (*5*) scirpane nomenclature[1], have now been reported from natural sources. They are made up of 133 (61%) non-macrocyclic and 84 (39%) macrocyclic compounds. Thus, 20 new non-macrocyclic trichothecenes have been isolated since 1995 and 17 new macrocycles since 1991.

Included in the total of non-macrocyclic trichothecenes are 35 (26%) trichoverroids (*6*) which have complex ester side chains at positions 4 and/or 15 of verrucarol (**2**; $R^1 = R^2 = R^5 = R^6 = H$, $R^3 = R^4 = OH$) and are biosynthetic precursors of, or shunt products from the biosynthesis of, the macrocycles. The remaining 98 non-macrocyclic compounds are designated "simple" trichothecenes. In the macrocycles two side chains of a trichoverroid are joined to form an 18-membered ring. As would be expected, the trichoverroids and the macrocycles are produced by the same organisms.

Also included in the 133 non-macrocyclic trichothecenes are one uncharacterised compound, seven compounds detected only in the mass

(**1**)

---

[1] Scirpane (= 12,13-epoxytrichothecane) nomenclature is still used in devising trivial names for new compounds.

References, pp. 113–130

spectrometer, and thus incompletely characterised, and three possible artefacts (see Notes to Tables 1–4). Included in the macrocycles are two possible artefacts (see Notes to Tables 5–7). If all the compounds about which there is some element of doubt are removed from the totals, the number of known naturally occurring trichothecenes is reduced to 204.

For tabulation, the non-macrocyclic trichothecenes have been subdivided into 12,13-epoxytrichothec-9-enes (103, 77%) (Table 1), 12,13-epoxytrichothec-9-en-8-ones (23, 17%) (Table 2), trichothec-9,12-dienes (5, 4%) (Table 3), and miscellaneous (2) (Table 4). Thus, all but five of the non-macrocyclic trichothecenes have the 12,13-epoxide group. The pair of diastereoisomers in Table 4 have a 9,10-epoxide and are the only non-macrocyclic compounds not to have the 9(10)-ene implicit in the name trichothecene.

Two non-macrocyclic trichothecenes, 8-deoxotrichothecinol A (Table 1: $C_{19}H_{26}O_5$), from *Holarrhena floribunda*, and miotoxin G (Table 1: $C_{29}H_{40}O_9$), from *Baccharis coridifolia*, are plant products (but see Notes to Tables 1–4).

Whilst the remaining non-macrocyclic trichothecenes are metabolic products of fungi, 36 macrocyclic trichothecenes (43%), the baccharinoids, have been isolated *only* from plants of the genus *Baccharis*, and are listed in Table 7. The remaining 48 macrocycles are fungal products, classified as verrucarins (12, 25%) [skeleton (**6**)] (Table 5), mainly C_{27} compounds; or roridins (36, 75%) [skeleton (**18**)] (Table 6), mainly C_{29} compounds. Among these macrocycles, verrucarins A and J, and roridins A, D, E, and H have also been obtained from *Baccharis* spp., as has the simple trichothecene diacetylverrucarol.

Table 1. Non-Macrocycles: 12,13-Epoxytrichothec-9-enes (**2**; $R^6 = H$)[a]

Formula	R^1	R^2	R^3	R^4	R^5	Trivial Name	Source[b,c]	References[d]
$C_{15}H_{22}O_2$	H	H	H	H	H	Scirpene (12,13-Epoxytrichothecene)	*T. roseum, F. culmorum, F. crookwellense*	10, 11, 12
							F. graminearum, **Spicellum roseum**	12, 13
$C_{15}H_{22}O_3$	H	H	H	OH	H	Trichodermol (Roridin C)	*M. roridum, Trichoderma polysporum*	14, 15
							S. cylindrospora, T. roseum	16, 15
							Memnoniella echinata, Spicellum roseum	17, 13
							Gliocladium virens	18
$C_{15}H_{22}O_3$	H	H	H	H	OH	Isotrichodermol	*F. crookwellense, F. graminearum*	19, 12
							F. venenatum	20
$C_{15}H_{22}O_3$	OH[e]	H	H	H	H	8-Hydroxyscirpene	*F. sporotrichioides*	21
$C_{15}H_{22}O_4$	OH	H	H	H	OH	8-Hydroxyisotrichodermol	*F. crookwellense*, **F. graminearum**	19, 22
							F. sporotrichioides	23
$C_{15}H_{22}O_4$	H	OH	H	H	OH	7-Hydroxyisotrichodermol	*F. crookwellense*, **F. graminearum**	19, 22
$C_{15}H_{22}O_4$	OH	H	H	OH	H	Trichothecodiol	*T. roseum*	10
$C_{15}H_{22}O_4$	H	OH	H	OH	H	7-Hydroxytrichodermol	*M. roridum*	24
$C_{15}H_{22}O_4$	H	H	OH	OH	H	Verrucarol[f]	*S. atra, S. microspora*, **F. sporotrichioides**	25, 25, 26
$C_{15}H_{22}O_4$	H	H	OH	H	OH	Isoverrucarol	*F. culmorum, F. sporotrichioides*	27, 28
							F. oxysporum	29
$C_{15}H_{22}O_5$	OH	H	H	OH	OH	**Scirpen-3,4,8-triol**	**F. sporotrichioides**	23
$C_{15}H_{22}O_5$	H	H	OH	OH	OH	Scirpentriol[f,g] (Scirpen-3,4,15-triol)	*F. semitectum (roseum), F. sporotrichioides*	30, 31
							F. sambucinum, F. camptoceras	32, 33
							F. acuminatum, F. equiseti, **F. poae**	34, 34, 35
							F. venenatum	36
$C_{15}H_{22}O_6$	OH	H	OH	OH	OH	T-2 tetraol[f]	*F. sporotrichioides, F. acuminatum*	37, 38
							(*heterosporum*), *F. acuminatum, F. poae*	34, 37

Formula						Name	Organism	Refs
$C_{15}H_{22}O_6$	OH	OH	H	OH		Scirpen-3,7,8,15-tetraol	F. graminearum	39
$C_{17}H_{24}O_4$	H	H	OAc	H		Trichodermin	Trichoderma viride, Trichoderma polysporum, S. cylindrospora, Dendrostilbella sp.	40, 15; 16, 41
							Memnoniella echinata, Gliocladium virens	17, 18
$C_{17}H_{24}O_4$	H	H	H	OAc		Isotrichodermin	F. graminearum (roseum), F. culmorum	42, 27
							F. crookwellense, F. sambucinum	19, 12
							F. sporotrichioides, F. venenatum	23, 20
$C_{17}H_{24}O_5$	OH	H	H	OAc		8-Hydroxyisotrichodermin	F. graminearum (roseum), F. crookwellense	43, 19
							F. culmorum, **F. sporotrichioides**	44, 23
$C_{17}H_{24}O_5$	H	OH	H	OAc		7-Hydroxyisotrichodermin	F. graminearum (roseum), F. crookwellense	43, 19
							F. culmorum	44
$C_{17}H_{24}O_5$	H	OH	H	OAc		15-Deacetylcalonectrin	F. culmorum (Calonectria nivalis)	45
							F. graminearum (roseum), F. sporotrichioides	42, 28
$C_{17}H_{24}O_5$	H	OAc	H	OH		3-Deacetylcalonectrin	F. culmorum, F. crookwellense, F. graminearum	27, 12, 12
$C_{17}H_{24}O_6$	H	OAc	OH	OH		15-Acetoxyscirpendiol	F. equiseti (avenaceum, concolor, semitectum)	46, 46, 46
							F. equiseti, F. sambucinum (sulphureum)	46, 47
							F. sambucinum, F. sporotrichioides	32, 48
							F. poae, **Cylindrocladium floridanum**	49, 50
							F. semitectum (roseum), **F. venenatum**	30, 36
$C_{17}H_{24}O_6$	H	OH	OAc	OH		4-Acetoxyscirpendiol[f]	F. semitectum (roseum), F. sambucinum	51
							(sulphureum), F. sambucinum, F. camptoceras	47, 32, 33
							Cylindrocladium floridanum, F. venenatum	50, 36
$C_{17}H_{24}O_6$	H	OH	OH	OAc		3-Acetoxyscirpendiol[f]	F. sambucinum, F. camptoceras	52, 33
$C_{17}H_{24}O_6$	OH	H	H	OAc		7,8-Dihydroxyisotrichodermin	F. crookwellense	19
$C_{17}H_{24}O_7$	OH	H	OAc	OH		15-AcetylT-2 tetraol	F. acuminatum (heterosporum)	38
							F. acuminatum, F. sporotrichioides	53, 54 (48[h])

Table 1 (continued)

Formula	R¹	R²	R³	R⁴	R⁵	Trivial Name	Source[b,c]	References[d]
$C_{17}H_{24}O_7$	OAc	H	OH	OH	OH	8-Acetyl[T-2 tetraol	*F. sporotrichioides, F. acuminatum*	*54 (48[h]), 53*
$C_{17}H_{24}O_7$	OH	OH	OH	H	OAc	7,8-Dihydroxy-15-deacetylcalonectrin	*F. culmorum*	*55*
$C_{17}H_{24}O_7$	OH	H	OH	OAc	OH	NT-2	*F. sporotrichioides (solani), F. acuminatum*	*56, 53*
$C_{19}H_{24}O_5$	—O—[i]		H	OR⁷	H	Crotocin	*Cephalosporium crotocinigenum, T. roseum*	*57, 58*
$C_{19}H_{26}O_4$	H	H	H	OR⁷	H	Isocrotonyltrichodermol (8-Deoxotrichothecin)	*T. roseum, Spicellum roseum*	*59, 60*
$C_{19}H_{26}O_5$	H	H	H	OR⁷	OH	**8-Deoxotrichothecinol A**[j]	***Holarrhena floribunda***[k]	*61*
$C_{19}H_{26}O_5$	OH	H	H	OR⁷	H	**Trichothecinol B**	***T. roseum***	*62*
$C_{19}H_{26}O_6$	OH	H	H	OR⁷	OH	**Trichothecinol C**	***T. roseum***	*62*
$C_{19}H_{26}O_6$	H	H	OAc	OAc	H	Diacetylverrucarol[f]	*M. verrucaria, B. coridifolia*[l]	*63, 64*
$C_{19}H_{26}O_6$	H	H	OAc	H	OAc	Calonectrin	*F. culmorum (Calonectria nivalis)*	*45*
							F. graminearum (roseum)	*43*
							F. sporotrichioides, F. crookwellense	*65, 12*
$C_{19}H_{26}O_7$	H	H	OH	OAc	OAc	3,4-Diacetoxyscirpenol[f]	*F. sambucinum*	*52*
$C_{19}H_{26}O_7$	H	H	OAc	OH	OAc	3,15-Diacetoxyscirpenol[f]	*F. sambucinum*	*32*
$C_{19}H_{26}O_7$	H	H	OAc	OAc	OH	Diacetoxyscirpenol (4,15-Diacetoxyscirpenol)	*F. equiseti, F. equiseti (avenaceum concolor, semitectum)*	*5, 46*
							F. equiseti (roseum), F. graminearum (tricinctum, roseum)	*46, 46*
							F. sporotrichioides (solani, tricinctum)	*66*
							F. sporotrichioides, F. sambucinum	*66, 66*
							(sulphureum, roseum), F. sambucinum	*66, 66*
								66
								47, 51, 67

$C_{19}H_{26}O_7$	OH	H	OAc	H	OAc	8-Hydroxycalonectrin
$C_{19}H_{26}O_7$	H	OH	OAc	H	OAc	7-Hydroxycalonectrin
$C_{19}H_{26}O_8$	OH	H	OAc	OAc	OH	Neosolaniol[m]

F. poae (tricinctum), F. graminearum — 68 (116), 69
F. acuminatum, F. lateritium, F. oxysporum — 9a, 70, 71
F. oxysporum (lateritium), F. solani — 72, 73
F. equiseti (compactum), F. moniliforme — 74, 75
F. culmorum, F. poae, F. avenaceum — 12, 76, 69
F. crookwellense, F. semitectum — 77, 78
F. venenatum — 79
F. culmorum, F. graminearum (roseum) — 27, 43
F. graminearum (roseum), F. culmorum — 43, 44
F. sporotrichioides (solani, poae, tricinctum) — 80, 82, 66
F. sporotrichioides, F. graminearum (decemcellulare, roseum) — 82
F. sambucinum (roseum), F. acuminatum (sulphureum), F. equiseti, F. sambucinum — 66, 66
F. avenaceum, F. semitectum — 51
F. acuminatum, F. solani, F. crookwellense — 83, 84, 85
F. tumidum, F. poae, F. oxysporum (lateritium) — 66, 78
— 78, 86, 87
— 88, 37
— 84

$C_{19}H_{26}O_8$	OAc	H	OH	OAc	OH	NT-1

F. sporotrichioides (solani, tricinctum) — 56, 89
F. sporotrichioides, F. acuminatum — 90, 53

$C_{19}H_{26}O_8$	OAc	H	OAc	OH	OH	Acuminatin[f]
$C_{19}H_{26}O_8$	OH	OH	OAc	H	OAc	7,8-Dihydroxycalonectrin
$C_{19}H_{26}O_8$	H	OH	OAc	OAc	OH	7-Hydroxydiacetoxyscirpenol

F. acuminatum, F. equiseti (compactum)[n] — 53, 74
F. graminearum (roseum), F. culmorum — 42[o], 91
F. oxysporum (lateritium), f. graminearum (roseum) — 84
— 92

$C_{19}H_{26}O_9$	OH	OH	OAc	OAc	OH	7,8-Dihydroxydiacetoxyscirpenol

F. oxysporum (lateritium), F. camptoceras — 84, 33

Table 1 (continued)

Formula	R^1	R^2	R^3	R^4	R^5	Trivial Name	Source[b,c]	References[d]
$C_{20}H_{30}O_6$	OR^8	H	OH	H	OH	Sporotrichiol	*F. sporotrichioides*	54
$C_{20}H_{30}O_7$	OR^8	H	OH	OH	OH	T-2 triol[f]	*F. sporotrichioides*	93
$C_{20}H_{30}O_7$	OH	H	OR^8	OH	OH	15-IsovalerylT-2 tetraol[h]	*F. sporotrichioides*	48
$C_{20}H_{30}O_8$	OR^9	H	OH	OH	OH	3'-HydroxyT-2 triol	*F. acuminatum (heterosporum)*	38
$C_{21}H_{28}O_8$	H	H	OAc	OAc	OAc	Triacetoxyscirpene[f]	*F. sambucinum (sulphureum)*	47
							F. sambucinum, **F. poae**	32, 35
$C_{21}H_{28}O_9$	OAc	H	OAc	OAc	OH	9-Acetylneosolaniol	*F. chlamydosporium (tricinctum)*	94
							F. equiseti (compactum)	74
							F. sambucinum (roseum)	51
							F. sambucinum, F. acuminatum	85, 53
							F. sporotrichioides. F. sp.	95, 96
							F. equiseti	5
$C_{21}H_{28}O_{10}$	OAc	OH	OAc	OAc	OH	Verrol	*M. verrucaria*, **S. atra**	97, 98
$C_{21}H_{30}O_6$	H	H	OR^{10}	OH	H		*F. sporotrichioides*	48
$C_{21}H_{30}O_8$[h]	OR^{11}	H	OAc	OH	OH		*F. sporotrichioides*	48
$C_{22}H_{30}O_8$[h,p]	OR^{12}	H	OAc	OH	OH		*F. sporotrichioides*	48
$C_{22}H_{30}O_9$	OR^{13}	H	OAc	OAc	OH	8-Propionylneosolaniol	*F. sporotrichioides, F. sambucinum*	65, 85
$C_{22}H_{32}O_7$	OR^8	H	OAc	H	OH	4-DeacetoxyT-2 toxin	*F. sporotrichioides*	90
$C_{22}H_{32}O_8$	OR^8	H	OH	OAc	OH	15-DeacetylT-2 toxin[h]	*F. sporotrichioides*	48
$C_{22}H_{32}O_8$	OR^8	H	OAc	OH	OH	HT-2 toxin	*F. sporotrichioides (tricinctum, solani, poae)*	99, 66, 100
							F. sporotrichioides	37
							F. graminearum (tricinctum, roseum)	66, 66
							F. acuminatum (sulphureum, heterosporum)	83, 38

The Trichothecenes and Their Biosynthesis

Formula	R	R	R	R	R	Compound	Organism	Ref.
$C_{22}H_{32}O_9$	OR^9	H	OAc	OH	OH	3'-HydroxyHT-2 toxin	*F. acuminatum, F. sambucinum*	86, 85
$C_{23}H_{28}O_6$	H	H	H	OR^{14}	H	Harzianum A	*F. oxysporum, F. equiseti, F. solani*	71, 86, 86
$C_{23}H_{30}O_5$	H	H	H	OR^{15}	H	Trichodermadiene	*F. culmorum, F. graminearum*	87, 87
$C_{23}H_{30}O_{10}$	OAc	H	OAc	OAc	OAc	Diacetylneosolaniol[f]	*F. moniliforme, F. poae*	101, 37
$C_{23}H_{32}O_6$	H	H	H	OR^{16}	H	Trichodermadienediol A,B[q]	*F. acuminatum (heterosporum)*	38
$C_{23}H_{32}O_7$	H	H	OR^{10}	OAc	H	**4-Acetylverrol**	*F. sporotrichioides*	102
$C_{23}H_{32}O_7$[r]	H	H	H	OR^{16}	H	16-Hydroxytrichodermadienediol A,B[q]	*Trichoderma harzianum*	103
$C_{23}H_{32}O_7$	H	H	OH	OR^{16}	H	Trichoverrol A,B[q]	*M. verrucaria*	104
$C_{23}H_{32}O_7$	H	H	OH	OR^{17}	H	Isotrichoverrol A,B[s]	*F. acuminatum*	53
$C_{23}H_{32}O_7$	H	H	OH	OR^{18}	H	**(2'E)-Isotrichoverrol A,B**[s]	*M. verrucaria, M. roridum, S. atra*	105, 106, 107
$C_{23}H_{32}O_9$	OR^{11}	H	OAc	OAc	OH	8-Butyrylneosolaniol	*Acremonium neo-caledoniae*	108
$C_{23}H_{32}O_9$	OR^{19}	H	OAc	OAc	OH	8-Isobutyrylneosolaniol	*M. verrucaria, M. roridum*	109, 109
$C_{23}H_{34}O_{11}$	H	H	OAc	OR^{20}	OH	15-Acetoxyscirpendiol-4-β-glucoside	*M. verrucaria, S. atra, S. albipes* *S. kampalensis, S. microspora*	105, 110, 111 111, 111
$C_{24}H_{32}O_9$	OR^{12}	H	OAc	OAc	OH	8-Pentenoylneosolaniol[h,t]	*M. verrucaria*	112
$C_{24}H_{34}O_9$	OR^{12}	H	OAc	OAc	OH	8-*n*-Pentanoylneosolaniol	*M. verrucaria*	113
$C_{24}H_{34}O_9$	OR^8	H	OAc	OAc	OH	T-2 toxin	*F. sporotrichioides, F. sambucinum* *F. sporotrichioides* *F. sambucinum (sulphureum)* *F. sporotrichioides* *F. sporotrichioides* *F. sporotrichioides (tricinctum, solani moniliforme, poae), F. sporotrichioides* *F. avenaceum (roseum), F. semitectum* *F. acuminatum (sulphureum, heterosporum)*	65 (48[h]), 85 65 114 48 115 116, 66 117, 100, 37 66, 118 83, 38

Table 1 (continued)

Formula	R^1	R^2	R^3	R^4	R^5	Trivial Name	Source[b,c]	References[d]
							F. equiseti, F. sambucinum (roseum)	118, 51
							F. graminearum (tricinctum, roseum)	66, 66
							F. culmorum, F. acuminatum, F. poae	69, 119, 120
							F. graminearum, F. oxysporum	69, 121
							F. moniliforme, F. sambucinum	75, 85
							F. crookwellense, F. stilboides, F. solani	69, 69, 86
							F. subglutinans	101
							Trichoderma viride (lignorum)	99
$C_{24}H_{34}O_9$	OR^8	H	OAc	OH	OAc	IsoT-2 toxin[f]	F. sporotrichioides, F. graminearum	71, 122
$C_{24}H_{34}O_{10}$	OR^9	H	OAc	OAc	OH	3'-HydroxyT-2 toxin	F. sporotrichioides, F. poae, F. oxysporum	102, 123, 71
$C_{25}H_{36}O_9$	OR^8	H	OAc	OR^{13}	OH	4-PropionylHT-2 toxin	F. sporotrichioides	21
$C_{25}H_{36}O_9$	OR^{22}	H	OAc	OAc	OH	8-n-Hexanoylneosolaniol	F. sporotrichioides	115
$C_{26}H_{36}O_{10}$	OR^8	H	OAc	OAc	OAc	AcetylT-2 toxin[f]	F. sporotrichioides (poae)	124
							F. sporotrichioides, F. graminearum	71, 122
$C_{29}H_{38}O_9$	H	H	OH	OR^{23}	H	Roridin L-2	M. roridum, S. atra	125, 107
$C_{29}H_{38}O_9$	H	H	OH	OR^{24}	H	**(2′E)-Roridin L-2**[u]	**M. verrucaria**	113
$C_{29}H_{38}O_{10}$[r]	H	H	OH	OR^{23}	H	16-Hydroxyroridin L-2	M. roridum	126
$C_{29}H_{40}O_9$	H	H	OH	OR^{25}	H	**Miotoxin G**[v]	**B. coridifolia**	127
$C_{29}H_{40}O_9$	H	H	OR^{10}	OR^{16}	H	Trichoverrin A,B[q]	M. verrucaria, S. atra	105, 110
$C_{29}H_{40}O_9$	H	H	OR^{10}	OR^{17}	H	**Isotrichoverrin A,B**[s]	**M. verrucaria**	113
$C_{29}H_{40}O_9$	H	H	OR^{10}	OR^{18}	H	(2″E)-Isotrichoverrin A,B[s]	M. verrucaria	112
$C_{29}H_{40}O_9$	H	H	OR^{10}	OR^{26}	H	**(2′E,4″Z)-Isotrichoverrin A,B**[s]	**M. verrucaria**	113
$C_{29}H_{40}O_9$	H	H	OR^{27}	OR^{17}	H	[iso]Trichoverrin C[w]	M. verrucaria	112
$C_{29}H_{40}O_9$	OH	H	OR^{10}	OR^{17}	H	**8-Hydroxyisotrichoverrin A**	**M. verrucaria**	113

$C_{31}H_{52}O_6$	H	H	OR^{28}	OR^{28}	H	Palmitylscirpentriol[h]	F. moniliforme	128
$C_{31}H_{52}O_7$	OR^{29}	H	OR^{29}	OR^{29}	H	PalmitylT-2 tetraol[h]	F. moniliforme	128
$C_{35}H_{46}O_{11}$	H	H	OR^{10}	OR^{23}	H	Trichoverritone	M. roridum	126

[a] New compounds and sources reported since Dec. 1995 are printed in bold type.
[b] B = *Baccharis*, F = *Fusarium*, M = *Myrothecium*, S = *Stachybotrys*, T = *Trichothecium*.
[c] Fusaria are named according to the scheme proposed by Booth (7), as extended and amended by Nelson *et al.* (8). Where reinvestigation has led to a change in identification (9), the original assignment is given in parentheses.
[d] The reference is to the first isolation from each source.
[e] 8β-OH.
[f] Known compound at time of isolation: previously obtained by chemical modification of other naturally-occurring trichothecenes.
[g] 3,4,15-Oxygenation is assumed unless stated otherwise.
[h] Ms identification only: not completely characterized.
[i] 7β,8β-Epoxide.
[j] Incorrectly named 8-dihydrotrichothecinol A.
[k] A West African shrub. The workers (61) could not exclude the possibility that this compound arose from fungal contamination of the plant material. The material tested negative for *Fusarium* spp.; however, the pattern of the *known* metabolic products isolated included the diterpenoids rosenonolactone, 6β-hydroxy-rosenonolactone, and rosololactone, and is typical of a *T. roseum* strain.
[l] The derivative, 4β,15-diacetoxy-10,13-cyclotrichothecan-9α,12-diol, $C_{19}H_{28}O_7$, also isolated from *B. coridifolia* (53) is presumed to be an artefact.
[m] Initially named solaniol (80), but later renamed neosolaniol (81).
[n] Initially believed to be the isomer NT1.
[o] The correct configuration for the 8-hydroxy substituent in this compound is α (54).
[p] Possible artefact by degradation of 3′-hydroxyHT-2 toxin.
[q] Diastereoisomers, epimeric at C-7′. Isomer A has C-7′S.
[r] R^6 = OH
[s] Diastereoisomers, epimeric at C-7′ The prefix "iso" indicates that C-6′ is R. In this series isomer A has C-7′R.
[t] Possible artefact by degradation of 3′-hydroxyT2 toxin.
[u] Claimed, but not characterized.
[v] Possible artefact derived from the macrocycle miotoxin D by hydrolysis and relactonisation.
[w] Named trichoverrin C, but on the "iso" convention (notes) should be isotrichoverrin C.

Radicals for Tables 1–4

$R^7 = CO \cdot CH \stackrel{Z}{=} CHMe$

$R^8 = CO \cdot CH_2 \cdot CHMe_2$

$R^9 = CO \cdot CH_2 \cdot C(OH)Me_2$

$R^{10} = CO \cdot CH \stackrel{E}{=} CMe \cdot CH_2 \cdot CH_2OH$

$R^{11} = CO \cdot C_3H_7$

$R^{12} = CO \cdot C_4H_7$

$R^{13} = CO \cdot Et$

$R^{14} = CO \cdot CH \stackrel{Z}{=} CH \cdot CH \stackrel{E}{=} CH \cdot CH \stackrel{E}{=} CH \cdot CO_2H$

$R^{15} = CO \cdot CH \stackrel{Z}{=} CH \cdot CH \stackrel{E}{=} CH \cdot \underset{\underset{O}{\diagup}}{\overset{H \; R}{C}} \underset{\overset{}{H}}{\overset{R}{-}} \overset{}{C}Me$

$R^{16} = CO \cdot CH \stackrel{Z}{=} CH \cdot CH \stackrel{E}{=} CH \cdot \overset{6' \; S}{CHOH} \cdot \overset{7' \; S,R}{CHOH} \cdot Me$

$R^{17} = CO \cdot CH \stackrel{Z}{=} CH \cdot CH \stackrel{E}{=} CH \cdot \overset{6' \; R}{CHOH} \cdot \overset{7' \; R,S}{CHOH} \cdot Me$

$R^{18} = CO \cdot CH \stackrel{E}{=} CH \cdot CH \stackrel{E}{=} CH \cdot \overset{6' \; R}{CHOH} \cdot \overset{7' \; R,S}{CHOH} \cdot Me$

$R^{19} = CO \cdot CHMe_2$

$R^{20} = C_6H_{11}O_5$

$R^{21} = CO \cdot C_4H_9$

$R^{22} = CO \cdot C_5H_{11}$

$R^{23} = CO \cdot CH \stackrel{Z}{=} CH \cdot CH \stackrel{E}{=} CH \cdot CH \cdot O \cdot CH_2 \cdot CH_2 C = CH$
  with MeCHOH branch and $CH_2 - O$, $=O$ ring

$R^{24} = CO \cdot CH \stackrel{E}{=} CH \cdot CH \stackrel{E}{=} CH \cdot CH \cdot O \cdot CH_2 \cdot CH_2 C = CH$
  with MeCHOH branch and $CH_2 - O$, $=O$ ring

$R^{25} = CO \cdot CH \stackrel{Z}{=} CH \cdot CH \stackrel{E}{=} CH \cdot \overset{R}{CH} \cdot O \cdot CH_2 \cdot \overset{H}{C} - CH \stackrel{Me}{}$
  with MeCHOH (R) and ring CH_2, O, $C=O$

$R^{26} = CO \cdot CH \stackrel{E}{=} CH \cdot CH \stackrel{Z}{=} CH \cdot \overset{6' \; R}{CHOH} \cdot \overset{7' \; R,S}{CHOH} \cdot Me$

$R^{27} = CO \cdot CH_2 \cdot CMe \stackrel{E}{=} CH \cdot CH_2OH$

$R^{28} = [H,H,CO(CH_2)_{14}Me]$

$R^{29} = [H,H,H,CO(CH_2)_{14}Me]$

$R^{30} = CO \cdot CHOH \cdot Me$

$R^{31} = CO \cdot CH=CH \cdot CH=CH \cdot Me$

$R^{32} = CO \cdot CH \stackrel{E}{=} CH \cdot Ph$

$R^{33} = CO \cdot (CH_2)_{14} Me$

References, pp. 113–130

Table 2. Non-Macrocycles: 12,13-Epoxytrichothec-9-en-8-ones (**3**)[a]

Formula	R^1	R^2	R^3	R^4	Trivial Name	Source[b,c]	References[d]
$C_{15}H_{20}O_4$	H	H	OH	H	Trichothecolone[e]	*T. roseum, F. moniliforme,* **Holarrhena floribunda**[g]	58, 128, 61
$C_{15}H_{20}O_5$	H	OH	H	OH	7-Deoxyvomitoxin	*F. graminearum*	129
$C_{15}H_{20}O_6$	OH	OH	H	OH	Vomitoxin (4-Deoxynivalenol)	*F. graminearum (roseum), F. graminearum*	130, 131
						F. culmorum, F. sporotrichioides	132
						(tricinctum, moniliforme) F. sporotrichioides	117, 117, 95
						F. solani, F. sambucinum, F. avenaceum	133, 133, 134
						F. oxysporum, F. equiseti, F. moniliforme	135, 135, 136
						F. semitectum, F. acuminatum, F. poae	134, 87, 87
						F. crookwellense, F. subglutinans	87, 137
						Microdochium nivale (F. nivale)	117
$C_{15}H_{20}O_6$	H	OH	OH	OH	7-Deoxynivalenol	*F. graminearum, F. camptoceras*	138, 33
$C_{15}H_{20}O_7$	OH	OH	OH	OH	Nivalenol	*F. sporotrichioides (nivale, episphaeria)*	139, 66
						F. sporotrichioides, F. semitectum, F. sambucinum	87, 78
						(sulphureum), F. equiseti, F. graminearum	78, 78, 66
						F. crookwellense, F. camptoceras, F. poae	140, 33, 141
						F. culmorum, F. solani, F. avenaceum	87, 142, 143
						F. sambucinum, F. oxysporum	143, 143
$C_{17}H_{22}O_5$	H	H	OAc	H	Acetyltrichothecolone[e]	*T. roseum*	144
$C_{17}H_{22}O_5$	H	H	H	OAc	8-Oxoisotrichodermin	*F. crookwellense*	19
$C_{17}H_{22}O_6$	H	OH	H	OAc	8-Oxo-15-deacetylcalonectrin	*F. graminearum (roseum), F. culmorum*	42, 27
$C_{17}H_{22}O_7$	OH	OH	H	OAc	3-Acetylvomitoxin	*F. culmorum, F. graminearum, F. graminearum*	145, 146
						(roseum), F. semitectum, F. camptoceras	130, 134, 33
						F. solani, F. acuminatum, F. avenaceum	133, 143, 143
						F. sambucinum, F. oxysporum	143, 143
$C_{17}H_{22}O_7$	OAc	OH	H	OH	7-Acetylvomitoxin	*F. camptoceras*	33
$C_{17}H_{22}O_7$	OH	OAc	H	OH	15-Acetylvomitoxin	*F. graminearum, F. sporotrichioides, F. semitectum*	147, 86, 134
						F. crookwellense, F. avenaceum, F. culmorum	87, 87, 87
						F. equiseti, F. poae	87, 87
$C_{17}H_{22}O_8$	OH	OH	OAc	OH	Fusarenone	*F. sporotrichioides (nivale, episphaeria, oxysporum)*	148, 66, 66

Table 2 (continued)

Formula	R^1	R^2	R^3	R^4	Trivial Name	Source[b,c]	References[d]
					(Fusarenone X)	*F. sporotrichioides, F. sambucinum (sulphureum)*	87, 78
						F. equiseti, F. semitectum, F. graminearum	78, 78, 66
						F. culmorum, F. crookwellense, F. poae	134, 140, 141
						F. camptoceras, F. acuminatum, F. sambucinum	33, 87, 134
						F. avenaceum	143
$C_{18}H_{24}O_8$	OH	OH	H	OR^{30}	CBD_2	Fungus infected barley	149
$C_{19}H_{24}O_5$	H	H	OR^7	H	Trichothecin	*T. roseum, F. graminearum, Holarrhena floribunda*[g]	150, 151, 61
$C_{19}H_{24}O_6$	H	H	OR^7	OH	**Trichothecinol A**	***T. roseum, Holarrhena floribunda***[g]	62, 61
$C_{19}H_{24}O_7$	H	OAc	H	OAc	8-Oxocalonectrin	*F. culmorum*	27
$C_{19}H_{24}O_8$	H	OAc	OAc	OH	8-Oxodiacetoxyscirpenol[e]	*F. sporotrichioides, F. crookwellense, F. culmorum*	152, 19, 12
$C_{19}H_{24}O_8$	OH	OAc	H	OAc	3,15-Diacetylvomitoxin[e]	*F. graminearum (roseum), F. culmorum*	153, 11
$C_{19}H_{24}O_9$	OH	OAc	OAc	OH	4,15-Diacetylnivalenol	*F. equiseti, F. culmorum, F. oxysporum (lateritium)*	5, 145, 84
						F. sporotrichioides (nivale, oxysporum)	154, 66
						F. crookwellense, F. camptoceras, F. sambucinum	19, 33, 12
						F. graminearum, **F. poae**	12, 35
$C_{21}H_{26}O_6$	H	H	OR^{31}	OH	**F-11703-1**	***Acremonium* sp.**	155
$C_{21}H_{26}O_7$	OH	H	OR^{31}	OH	**F-11703-2**	***Acremonium* sp.**	155
$C_{24}H_{26}O_5$	H	H	OR^{32}	H	Cinnamyltrichothecolone	*T. roseum*	144
$C_{31}H_{50}O_5$	H	H	OR^{33}	H	Palmityltrichothecolone[f]	*F. moniliforme*	128

[a] New compounds and sources reported since Dec. 1995 are printed in bold type.
[b] B = *Baccharis*, F = *Fusarium*, M = *Myrothecium*, S = *Stachybotrys*, T = *Trichothecium*.
[c] Fusaria are named according to the scheme proposed by Booth (7), as extended and amended by Nelson *et al.* (8). Where reinvestigation has led to a change in identification (9), the original assignment is given in parentheses.
[d] The reference is to the first isolation from each source.
[e] Known compound at time of isolation: previously obtained by chemical modification of other naturally-occurring trichothecenes.
[f] Ms identification only; not completely characterized.
[g] A West African shrub. The workers (61) could not exclude the possibility that this compound arose from fungal contamination of the plant material. The material tested negative for *Fusarium* spp.; however, the pattern of the *known* metabolic products isolated included the diterpenoids rosenonolactone, 6β-hydroxy-rosenonolactone, and rosololactone, and is typical of a *T. roseum* strain.

Table 3. Non-Macrocylces: Trichothec-9,12-dienes (**4**)[a]

Formula	R^1	R^2	R^3	R^4	R^5	Trivial Name	Source[b,c]	References[d]
$C_{19}H_{26}O_6$	H	H	OAc	OAc	OH	12,13-Deoxydiacetoxyscirpenol	*F. graminearum*	*156*
$C_{23}H_{30}O_4$	H	H	H	OR^{15}	H	12,13-Deoxytrichodermadiene	*M. verrucaria*	*97*
$C_{29}H_{40}O_8$	H	H	OR^{10}	OR^{16}	H	12,13-Deoxytrichoverrin A,B[e]	*M. verrucaria*	*157*
$C_{29}H_{40}O_8$	H	H	OR^{10}	OR^{18}	H	**12,13-Deoxy-(2″E)-isotrichoverrin B**	*M. verrucaria*	*113*

[a] New compounds and sources reported since Dec. 1995 are printed in bold type.
[b] B = *Baccharis*, F = *Fusarium*, M = *Myrothecium*, S = *Stachybotrys*, T = *Trichothecium*.
[c] Fusaria are named according to the scheme proposed by Booth (*7*), as extended and amended by Nelson *et al.* (*8*). Where reinvestigation has led to a change in identification (*9*), the original assignment is given in parentheses.
[d] The reference is to the first isolation from each source.
[e] Diastereoisomers, epimeric at C-7′. Isomer A has C-7′S.

Table 4. Non-Macrocycles: Miscellaneous (5)[a]

Formula	R^1	R^2	R^3	R^4	R^5	Trivial Name	Source[b,c]	References[d]
$C_{29}H_{40}O_{10}$	H	H	OR^{10}	OR^{17}	H	**9β,10β-Epoxyisotrichoverrin A,B**[e]	*M. verrucaria*	*113*

[a] New compounds and sources reported since Dec. 1995 are printed in bold type.
[b] B = *Baccharis*, F = *Fusarium*, M = *Myrothecium*, S = *Stachybotrys*, T = *Trichothecium*.
[c] Fusaria are named according to the scheme proposed by Booth (7), as extended and amended by Nelson *et al.* (8). Where reinvestigation has led to a change in identification (9), the original assignment is given in parentheses.
[d] The reference is to the first isolation from each source.
[e] Diastereoisomers, epimeric at C-7' The prefix "iso" indicates that C-6' is R. In this series isomer A has C-7'R.

(40)

(39)

(38)

Table 5. Macrocycles: Verrucarins and Myrotoxins[a]

Formula	Trivial Name (synonym)	Structure	Source[b]	References[c]
$C_{27}H_{32}O_8$	Verrucarin J (Muconomycin B)[161] (Satratoxin C)[163]	(**6**; R = H)	*M. verrucaria*, *M. roridum* S. atra, S. kampalensis, S. albipes S. microspora, B. coridifolia, **B. artemisioides** **Ceratopycnidium baccharidicola** Unidentified marine fungus	*158, 159* *160, 111, 111* *25, 162, 164* *164* *165*
$C_{27}H_{32}O_9$	2′-Dehydroverrucarin A	(**7**; $R^1R^2 = O$)	*M. roridum*	*166*
$C_{27}H_{32}O_9$	PD 113325[d]	(**8**)	*M. roridum*	*159*
$C_{27}H_{32}O_9$	Myrotoxin A	(**9**; $R^1 = R^2 = H$, $R^3 = OH$)	*M. roridum*	*168*
$C_{27}H_{32}O_9$	Myrotoxin C	(**9**; $R^1 = R^3 = H$, $R^2 = OH$)	*M. roridum*	*169*
$C_{27}H_{32}O_9$	Verrucarin B (SIPI-299-O)[171]	(**10**)	*M. verrucaria, M. roridum, S. atra* **Phoma sp.**	*14, 14, 170* *171*
$C_{27}H_{32}O_9$	Verrucarin L	(**6**; R = OH)	*M. verrucaria*	*172*
$C_{27}H_{34}O_8$	Verrucarin K[e]	(**11**)	*M. verrucaria*	*175*
$C_{27}H_{34}O_9$	Verrucarin A (Antibiotic Y379)[177] (Muconomycin A)[178] (SIPI-299-B)[171]	(**7**; $R^1 = H$; $R^2 = OH$)	*M. verrucaria, M. roridum, M. leucotrichum,* *B. coridifolia,* **B. artemisioides** **Ceratopycnidium baccharidicola** *Acremonium neo-caledoniae,* **Phoma sp.**	*14, 14, 176* *162, 164* *164* *108, 171*
$C_{29}H_{34}O_{10}$	Acetylverrucarin L	(**6**; R = OAc)	*M. verrucaria,* **Unidentified marine fungus**	*172, 165*
$C_{29}H_{34}O_{11}$	Myrotoxin B	(**9**; $R^1 = OAc$, $R^2 = H$, $R^3 = OH$)	*M. roridum*	*168*
$C_{29}H_{34}O_{11}$	Myrotoxin D	(**9**; $R^1 = OAc$, $R^2 = OH$, $R^3 = H$)	*M. roridum*	*169*

[a] New compounds and sources reported since Dec. 1991 are printed in bold type.
[b] B = *Baccharis*, C = *Cylindrocarpon*, M = *Myrothecium*, S = *Stachybotrys*.
[c] The reference is to the first isolation from each source.
[d] Originally (*167*), incorrectly, named 12′-hydroxyverrucarin J.
[e] An earlier (*173*) bearer of this name was renamed roridin E (*174*).

Table 6. Macrocycles: Roridins and Roridin relatives[a]

Formula	Trivial Name (synonym)	Structure	Source[b]	References[c]
$C_{29}H_{32}O_{11}$	Roritoxin D	(12; R^1R^2 = O)	M. roridum	179
$C_{29}H_{32}O_{12}$	Roritoxin C	(13)	M. roridum	179
$C_{29}H_{34}O_9$	7β,8β-Epoxyroridin H	(14; R^1R^2 = O)	C. sp.	180
$C_{29}H_{34}O_{10}$	Diepoxyroridin H	(15)	C. sp.	180
$C_{29}H_{34}O_{10}$	Roritoxin A	(16)	M. roridum	179
$C_{29}H_{34}O_{10}$	Satratoxin F[d]	(17; R = COMe)	S. atra, S. kampalensis	181, 111
$C_{29}H_{34}O_{10}$	**Isosatratoxin F**[d]	(17; R = COMe)	**S. atra**	98
$C_{29}H_{34}O_{11}$	Roritoxin B	(12; R^1R^2 = H, OH)	M. roridum	179
$C_{29}H_{36}O_8$	Roridin H[e]	(14; $R^1 = R^2$ = H)	M. verrucaria, C. sp., **B. coridifolia**	158, 180, 164
$C_{29}H_{36}O_9$	7β,8β-Epoxyisororidin E	(18; R^1R^2 = O, $R^3 = \alpha H$)	C. sp.	180
$C_{29}H_{36}O_9$	Mytoxin B	(19)	M. roridum	169
$C_{29}H_{36}O_9$	Roridin J	(20)	M. verrucaria	182
$C_{29}H_{36}O_9$	Satratoxin H	(21; R^1 = H, R^2 = OH, C-13' S)	S. atra, S. kampalensis, S. microspora	163, 111, 25
$C_{29}H_{36}O_9$	PD 113326 (M Isosatratoxin H)[167, 107]	(21; R^1 = H, R^2 = OH, C-13' R)	M. roridum	159
$C_{29}H_{36}O_9$	Satratoxin H isomer[f] (S Isosatratoxin H)[107]	(21; R^1 = OH, R^2 = H)	S. atra	183
$C_{29}H_{36}O_9$	Satratoxin H isomer[f]	(21; R^1 = OH, R^2 = H)	S. atra	160
$C_{29}H_{36}O_{10}$	Mytoxin A	(22; R^1 = H, R^2 = OH)	M. roridum	169
$C_{29}H_{36}O_{10}$	Mytoxin C	(22; R^1 = OH, R^2 = H)	M. roridum	169
$C_{29}H_{36}O_{10}$	Satratoxin G[g]	(17; R = CHOH·Me) at 13'	S. atra, S. kampalensis	181, 111
$C_{29}H_{36}O_{10}$	**Isosatratoxin G**[h]	(17; R = CHOH·Me) at 13'	**S. atra**	107
$C_{29}H_{36}O_{10}$	Vertisporin	(23)	Verticinimonosporium diffractum	184
$C_{29}H_{38}O_7$	**12,13-Deoxyroridin E**	(24)	**M. roridum**	185

Table 6 (continued)

Formula	Trivial Name (synonym)	Structure	Source[b]	References[c]
$C_{29}H_{38}O_8$	Roridin E (Satratoxin D)[186]	(**25**; $R^1 = R^3 = H$, $R^2 = \beta H$)	*M. verrucaria, M. roridum, S. atra* *S. kampalensis, S. microspora* *B. coridifolia, B. megapotamica* **B. artemisioides**, *Cercophora areolata* *Ceratopycnidium baccharidicola*	158, 106, 160 111, 25 187, 188 164, 189 164
$C_{29}H_{38}O_8$	**13'-Epiroridin E**	(**25**; $R^1 = R^3 = H$, $R^2 = \alpha H$)	*S. atra, M. verrucaria*	107, 190
$C_{29}H_{38}O_8$	Isororidin E	(**18**; $R^1 = R^2 = H$, $R^3 = \alpha H$)	*C.* sp., *M. verrucaria, S. atra*	180, 6, 107
$C_{29}H_{38}O_8$	**13'-Epiisororidin E**	(**18**; $R^1 = R^2 = H$, $R^3 = \beta H$)	*S. atra, M. verrucaria*	107, 190
$C_{29}H_{38}O_8$	Roridin E-2[j]		*M. verrucaria*	191
$C_{29}H_{38}O_9$	8α-Hydroxyisororidin E[j] (**Isororidin K**)[190]	(**26**; $R^1 = H$, $R^2 = R^3 = \alpha H$)	*M.* sp., **M. Verrucaria**	192, 190
$C_{29}H_{38}O_9$	Roridin D	(**27**; $R = H$)	*M. roridum, B. megapotamica* *B. coridifolia,* **B. artemisioides** *Ceratopycnidium baccharidicola*	158, 188 193, 164 164
$C_{29}H_{40}O_9$	Roridin A (Antibiotic X379)[177]	(**28**; $R^1 = R^4 = H$, $R^2 = R^3 = \beta H$)	*M. roridum, M. verrucaria* *B. coridifolia, B. megapotamica* *Phomopsis leptostromiformis* **B. artemisioides** *Ceratopycnidium baccharidicola*	14, 14 187, 194 195 164 164
$C_{29}H_{40}O_9$	Isororidin A	(**28**; $R^1 = R^4 = H$, $R^2 = \beta H$, $R^3 = \alpha H$)	*M. verrucaria, Acremonium neo-caledoniae*	196, 108
$C_{29}H_{40}O_9$	**6'-Epi-13'-epiroridin A**[k]	(**28**; $R^1 = R^4 = H$, $R^2 = R^3 = \alpha H$)	*M.* sp., **M. verrucaria**	192, 190
$C_{31}H_{40}O_9$	**Acetylroridin E**	(**25**; $R^1 = Ac$, $R^2 = \beta H$, $R^3 = H$)	*M. verrucaria*	197

$C_{31}H_{40}O_{10}$	Acetylroridin K	(**26**; R^1 = Ac, $R^2 = R^3 = \beta H$)	*M. verrucaria*	6
$C_{33}H_{44}O_{11}$	**YM-47524**	(**28**; R^1 = OCOCH=CHMe, $R^2 = \beta H$, $R^3 = R^4 = H$)	**Unidentified fungus**	*198*
$C_{33}H_{46}O_{11}$	**YM-47525**	(**28**; R^1 = OCOCH$_2$CH$_2$Me, $R^2 = \beta H$, $R^3 = R^4 = H$)	**Unidentified fungus**	*198*

[a] New compounds and sources reported since Dec. 1991 are printed in bold type.
[b] B = *Baccharis*, C = *Cylindrocarpon*, M = *Myrothecium*, S = *Stachybotrys*.
[c] The reference is to the first isolation from each source.
[d] Likely to be epimeric at C-12′, but the nmr evidence was inconclusive (*98*).
[e] Previously verrucarin H (*158*).
[f] Possibly C-13′ epimers.
[g] C-12′-αOH: the configuration at C-13′ is unknown.
[h] Nmr evidence suggests a C-12′-epimer: the configuration at C-13′ is unknown.
[i] Well characterized isomer of roridin E. The 2′-ene may have the Z configuration (*191*).
[j] Originally (*192*) believed to be 8β-hydroxyroridin E.
[k] Originally (*192*) believed to be roridin A (6′R, 13′R), but is now considered (*190*), on compelling nmr evidence, to be the (6′S, 13′S) isomer.

Table 7. Macrocycles: Baccharinoids[a]

Formula	Trivial Name (synonym)	Structure	Source[b]	Refs.[c]
$C_{27}H_{32}O_{10}$	Baccharinoid B25[d]	(29)	B. megapotamica	199
$C_{29}H_{36}O_{10}$	Baccharinoid B27	(30)	B. megapotamica	199
$C_{29}H_{38}O_{8}$	Miophytocen A[e]	(31)	B. coridifolia	200
$C_{29}H_{38}O_{8}$	Miophytocen B[e]	(32)	B. coridifolia	200
$C_{29}H_{38}O_{9}$	Miotoxin A	(25; $R^1 = H, R^2 = \beta H, R^3 = OH$)	B. coridifolia	201
$C_{29}H_{38}O_{9}$	Miotoxin B	(33; $R^1 = R^2 = R^3 = R^6 = H, R^4R^5 = O$)	B. coridifolia	202
$C_{29}H_{38}O_{10}$	**Miotoxin E**	(33; $R^1 = R^3 = \beta H, R^2 = H, R^4R^5 = O, R^6 = OH$)	B. coridifolia	127
$C_{29}H_{38}O_{10}$	Baccharinoid B9	(34; $R = \beta H$)	B. megapotamica	199
$C_{29}H_{28}O_{10}$	Baccharinoid B10	(34; $R = \alpha H$)	B. megapotamica	199
$C_{29}H_{38}O_{10}$	Baccharinoid B12	(35; $R^1 = \beta H, R^2 = R^3 = H, R^4 = OH$)	B. megapotamica	199
$C_{29}H_{38}O_{10}$	Baccharinoid B13	(36; $R^1 = \beta H, R^2 = H, R^3 = OH$)	B. megapotamica	199
$C_{29}H_{38}O_{10}$	Baccharinoid B14	(36; $R^1 = \alpha H, R^2 = H, R^3 = OH$)	B. megapotamica	199
$C_{29}H_{38}O_{10}$	Baccharinoid B16	(36; $R^1 = \alpha H, R^2 = OH, R^3 = H$)	B. megapotamica	199
$C_{29}H_{38}O_{10}$	Baccharinoid B17	(37; $R^1 = \beta H, R^2 = H$)	B. megapotamica	199
$C_{29}H_{38}O_{10}$	Baccharinoid B21	(35; $R^1 = \beta H, R^2 = R^4 = H, R^3 = OH$)	B. megapotamica	199
$C_{29}H_{38}O_{11}$	Baccharinoid B4 (Baccharinol)	(35; $R^1 = \alpha H, R^2 = R^3 = OH, R^4 = H$)	B. megapotamica	203
$C_{29}H_{38}O_{11}$	Baccharinoid B5 (Baccharin)	(37; $R^1 = \beta H, R^2 = OH$)	B. megapotamica	204
			B. coridifolia (?)	193
$C_{29}H_{38}O_{11}$	Baccharinoid B6 (isoBaccharinol)	(35; $R^1 = \beta H, R^2 = R^3 = OH, R^4 = H$)	B. megapotamica	203
$C_{29}H_{38}O_{11}$	Baccharinoid B8 (isoBaccharin)	(37; $R^1 = \alpha H, R^2 = OH$)	B. megapotamica	203
$C_{29}H_{40}O_{9}$	Miotoxin D[f]	(33; $R^1 = R^3 = \beta H, R^2 = R^5 = R^6 = H, R^4 = OH$)	B. coridifolia	205
$C_{29}H_{40}O_{9}$	isoMiotoxin D[f]	(33; $R^1 = R^3 = \beta H, R^2 = R^5 = R^6 = H, R^4 = OH$)	B. coridifolia	205

Formula	Name	Structure	Source	Ref
$C_{29}H_{40}O_{10}$	**Miotoxin F**	(**33**; $R^1 = R^3 = \beta H$, $R^2 = H$, $R^4R^5 = H$, OH, $R^6 = OH$)	**B. coridifolia**	*127*
$C_{29}H_{40}O_{10}$	Baccharinoid B1	(**38**; $R^1 = \beta H$, $R^2 = R^4 = OH$, $R^3 = H$)	B. megapotamica	206
$C_{29}H_{40}O_{10}$	Baccharinoid B2	(**38**; $R^1 = \alpha H$, $R^2 = R^4 = OH$, $R^3 = H$)	B. megapotamica	206
$C_{29}H_{40}O_{10}$	Baccharinoid B3 (Baccharisol)	(**38**; $R^1 = \alpha H$, $R^2 = H$, $R^3 = R^4 = OH$)	B. megapotamica	206
			B. coridifolia	162
$C_{29}H_{40}O_{10}$	Baccharinoid B7 (isoBaccharisol)	(**38**; $R^1 = \beta H$, $R^2 = H$, $R^3 = R^4 = OH$)	B. megapotamica	206
			B. coridifolia	162
$C_{29}H_{40}O_{10}$	Baccharinoid B20	(**39**)	B. megapotamica	199
$C_{29}H_{40}O_{10}$	Baccharinoid B23	(**40**; $R = \beta H$)	B. megapotamica	199
$C_{29}H_{40}O_{10}$	Baccharinoid B24	(**40**; $R = \alpha H$)	B. megapotamica	199
$C_{31}H_{42}O_{11}$	Miotoxin C	(**33**; $R^1 = R^3 = \beta H$, $R^2 = Ac$, $R^4R^5 = H$, OH, $R^6 = OH$)	B. coridifolia	202
$C_{33}H_{44}O_{14}$	**Verrucarin A glucoside**	(**7**; $R^1 = H$, $R^2 = OC_6H_{11}O_5$)	**B. coridifolia**	*127*
$C_{35}H_{48}O_{13}$	**Roridin E glucoside**	(**25**; $R^1 = C_6H_{11}O_5$, $R^2 = \beta H$, $R^3 = H$)	**B. coridifolia**	*127*
$C_{35}H_{48}O_{14}$	**Roridin D glucoside**	(**27**; $R = C_6H_{11}O_5$)	**B. coridifolia**	*127*
$C_{35}H_{48}O_{14}$	**Miotoxin A 13′-glucoside**	(**25**; $R^1 = C_6H_{11}O_5$, $R^2 = \beta H$, $R^3 = OH$)	**B. coridifolia**	*127*
$C_{35}H_{50}O_{14}$	**Roridin A glucoside**	(**28**; $R^1 = H$, $R^2 = R^3 = \beta H$, $R^4 = C_6H_{11}O_5$)	**B. coridifolia**	*127*
$C_{35}H_{50}O_{15}$	**Miotoxin F glucoside**	(**33**; $R^1 = R^3 = \beta H$, $R^2 = C_6H_{11}O_5$, $R^4R^5 = H$, OH, $R^6 = OH$)	**B. coridifolia**	*127*

[a] New compounds and sources reported since Dec. 1991 are printed in bold type.
[b] B = *Baccharis*, C = *Cylindrocarpon*, M = *Myrothecium*, S = *Stachybotrys*.
[c] The reference is to the first isolation from each source.
[d] No evidence is presented (*199*) for an (E)-2′-ene. The chemical shift for C-12′ is consistent with the (Z) configuration.
[e] Possible artefact, derived from roridin E.
[f] C-3′ epimers.

The carbon skeleton (**9**) of the myrotoxins can formally be constructed from the verrucarin skeleton by formation of a C6′–C12′ bond, and this group of compounds is tabulated with the verrucarins. Likewise, the skeletons of the satratoxins, roritoxins, mytoxins and vertisporin can be obtained from the roridin skeleton by C6′–C12′ bond formation, and these compounds are tabulated with the roridins.

With the exception of verrucarin K (**11**) and 12,13-deepoxy-roridin E (**24**), which are 9(10),12(13)-dienes; and the miophytocens, which are 10,13-cyclo-trichothecanes, all the macrocycles have a 12,13-epoxide. The 9(10)-ene is replaced by a 9,10-epoxide in roritoxin C and in several baccharinoids.

2.2. Trichothecene Relatives

Some large scale and/or blocked fermentations with *Trichothecium roseum* or *Fusarium* spp. have yielded a number of metabolic products with structures closely related to the trichothecenes. These relatives, 49 in number (including 2 likely artefacts), an addition of six since 1995, are listed in Table 8 (structures: Scheme 1), together with their sources. They consist mainly of

(a) compounds with the tricho-9-ene skeleton (**41**)[2], some of which are intermediates in the biosynthesis of the trichothecenes, and
(b) from *Fusarium* spp., compounds with the 11-epiapo-trichothecene nucleus (**42**).

The *Fusarium* spp. metabolic products sambucinol (**81**) and its derivatives diacetylsambucinol (**82**) and 3-deoxysambucinol (**80**) can be regarded as 11-epi-12-epitrichothecenes but are more conveniently classified with the trichothecene relatives (Table 8), as is gramilaurone (**98**). These compounds are excluded from Tables 1–4 and the total of trichothecenes, as is the "isotrichothecin" (*233*) whose ^{13}C-nmr spectrum is indistinguishable from that of trichothecin.

---

[2] Biosynthetic numbering: no position 1.

References, pp. 113–130

Table 8. Naturally occurring trichothecene relatives[a]

Formula	Trivial Name	Structure	Source[b]	Refs.
$C_{15}H_{20}O_3$	FS 3	76	F. sambucinum	32
$C_{15}H_{20}O_4$	**Loukacinol B**	89	**Holarrhena floribunda**[c]	61
$C_{15}H_{20}O_5$	**Loukacinol A**	90	**Holarrhena floribunda**[c]	61
$C_{15}H_{20}O_2$		57	F. culmorum	207
$C_{15}H_{22}O_3$	FS 1	75	F. sporotrichioides, F. sambucinum	21, 32
$C_{15}H_{22}O_3$	FS 4	67[d]	F. sambucinum	32
$C_{15}H_{22}O_3$	3-Deoxysambucinol	80	F. culmorum, F. graminearum, F. crookwellense	208, 208, 19
$C_{15}H_{22}O_3$	Sambucinic acid	46	F. sambucinum	209
$C_{15}H_{22}O_3$	Sambucoin	69	F. sambucinum, F. sporotrichioides, F. poae	210, 65, 12
			F. culmorum, F. graminearum, F. crookwellense	27, 43, 19
$C_{15}H_{22}O_3$		92	F. sporotrichioides, F. culmorum, **F. graminearum**	211, 211, 22
$C_{15}H_{22}O_3$	3-Dehydroapotrichodiol	88	F. sporotrichioides, F. sambucinum	32, 32
$C_{15}H_{22}O_4$		93	F. sporotrichioides, **F. graminearum**	90, 22
$C_{15}H_{22}O_4$	8β-Hydroxysambucoin	70	F. sporotrichioides	212
$C_{15}H_{22}O_4$	8α-Hydroxysambucoin	71	F. sporotrichioides	212
$C_{15}H_{22}O_4$	Sambucinol	81	F. sambucinum, F. sporotrichioides, F. culmorum	210, 65, 27
			F. graminearum, F. crookwellense, **F. venenatum**	208, 19, 213
$C_{15}H_{22}O_4$	Sporol	91[e]	F. sporotrichioides	54 (214)
$C_{15}H_{22}O_7$	Gramilaurone	98[f]	F. graminearum	215
$C_{15}H_{24}$	Trichodiene	49	T. roseum, F. sambucinum, F. sporotrichioides	216, 217, 218
			F. culmorum, Monascus purpureus, **S. atra**	219, 220, 221
$C_{15}H_{24}O$	**2α-Hydroxytrichodiene**	47	**F. culmorum**	222
$C_{15}H_{24}O$	11α-Hydroxytrichodiene	48	F. sporotrichioides	223

Table 8 (continued)

Formula	Trivial Name	Structure	Source[b]	Refs.
$C_{15}H_{24}O$	**16-Hydroxytrichodiene**[g]	**44**	***Nicotiana tabacum***[h]	224
$C_{15}H_{24}O_2$	Isotrichool	52	*F. culmorum*	225
$C_{15}H_{24}O_2$	Apotrichool	85	*F. culmorum*, **F. sporotrichioides**	226, 26
$C_{15}H_{24}O_3$	Apotrichodiol	86[i]	*F. culmorum, F. graminearum, F. crookwellense*	208, 208, 19
			F. sporotrichioides, F. sambucinum, **F. venenatum**	65, 32, 20
$C_{15}H_{24}O_3$	3-epi-Apotrichodiol	87[j]	*F. culmorum, F. graminearum, F. crookwellense*	208, 208, 19
			F. sporotrichioides, F. sambucinum	65, 32
$C_{15}H_{24}O_3$		83	**F. culmorum**	228
$C_{15}H_{24}O_3$		100	**F. culmorum**	228
$C_{15}H_{24}O_3$		78	*F. sporotrichioides*	229
$C_{15}H_{24}O_3$	FS 2	65[d]	*F. sporotrichioides*	152
$C_{15}H_{24}O_3$	3-epi-FS 2	66	*F. sporotrichioides*	32
$C_{15}H_{24}O_3$	Trichodiol[j]	55[k,l]	*T. roseum*	230
$C_{15}H_{24}O_3$	9-epi-Trichodiol[m]	56	*T. roseum*	55
$C_{15}H_{24}O_3$	Isotrichodiol	53[k]	*F. culmorum*	219
$C_{15}H_{24}O_3$		51[n]	*F. sporotrichioides*	152 (231)
$C_{15}H_{24}O_4$	Trichotriol	63[k,o]	*F. sporotrichioides, F. culmorum*	152, 231
$C_{15}H_{24}O_4$	9-epi-Trichotriol[p]	64	*F. sporotrichioides, F. culmorum*	232, 231
$C_{15}H_{24}O_4$	Isotrichotriol	59[k]	*F. sporotrichioides, F. culmorum*	223, 55
$C_{15}H_{24}O_4$	8α-Hydroxyisotrichodiol	54	*F. culmorum*	231
$C_{15}H_{24}O_5$	8β-Hydroxyisotrichotriol	60	*F. sporotrichioides*	223
$C_{15}H_{24}O_5$	8α-Hydroxyisotrichotriol	61	*F. sporotrichioides*	223
$C_{15}H_{24}O_5$	16-Hydroxyisotrichotriol	62	*F. sporotrichioides*	223
$C_{15}H_{26}O_2$		43[q]	*F. culmorum*	225

References, pp. 113–130

$C_{17}H_{24}O_4$	AcetylFS 4	**68**[d]	*F. sambucinum*	207
$C_{17}H_{26}O_4$		**79**	*F. sporotrichioides*	229
$C_{17}H_{26}O_5$		**94**	*F. sporotrichioides*	90
$C_{19}H_{26}O_6$	Diacetylsambucinol	**82**	*F. sporotrichioides*	90
$C_{19}H_{28}O_6$		**95**	*F. sporotrichioides*	90
$C_{19}H_{28}O_6$		**96**	*F. sporotrichioides*	90

[a] New compounds and sources reported since December 1995 are printed in bold type.
[b] F = *Fusarium*, S = *Stachybotrys*, T = *Trichothecium*.
[c] A West African shrub. The workers (61) could not exclude the possibility that this compound arose from fungal contamination of the plant material. The material tested negative for *Fusarium* spp.; however, the pattern of the *known* metabolic products isolated included the diterpenoids rosenonolactone, 6β-hydroxy-rosenonolactone, and rosololactone, and is typical of a *T. roseum* strain.
[d] Originally formulated (152, 32, 207) as the 9α-hydroxy epimer, but changed (231) on spectroscopic evidence.
[e] An earlier, incorrect, structure (54) has been renamed neosporol.
[f] As published, but a possible artefact. The analogous structure (**99**) could be derived from vomitoxin (**3**; $R^1 = R^2 = R^4 = OH$, $R^3 = H$) (3).
[g] Reported (224) as 15-hydroxytrichodiene (distinct numbering).
[h] Cell suspension culture of *N. tabaccum* transformed with a gene encoding trichodiene synthase from *F. sporotrichioides*.
[i] Initially (208) formulated as an apotrichothecene, but subsequently shown (227, 211) to belong to the 11-epi series.
[j] Trichodiol A (**97**) is considered (229) to be an artefact.
[k] Some groups of workers, e.g. (225, 219), consistently (and mistakenly) write this compound as the 12-epimer.
[l] Originally (230) formulated without assignment of configuration at position 9, but commonly written as the 9α-hydroxy compound. It was subsequently shown to be the 9β-hydroxy epimer (231).
[m] Originally (55) called 9β-trichodiol.
[n] The metabolite was believed (152) to be trichodiol. This conclusion has been criticised (231), but the metabolite does not have the newly assigned (231) 12-ene structure: It is most probably (**51**).
[o] Initially (152) written as the 9α-hydroxy epimer by analogy with the commonly written structure for trichodiol.
[p] Originally (232) called 9β-trichotriol.
[q] Or C-12 epimer.

Scheme 1 (*continued*)

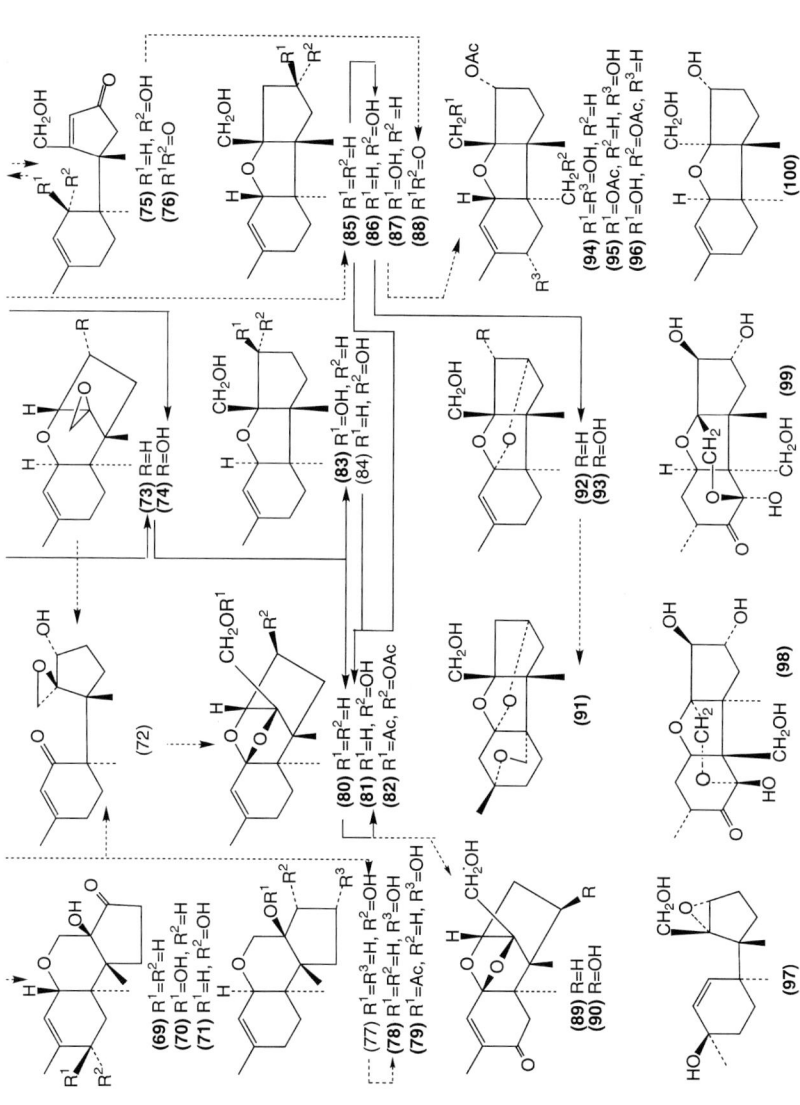

Scheme 1. Proven (——) and hypothetical (- - -) pathways from trichodiene (**49**) to the trichothecene relatives listed in Table 8. (nn) Isolated trichothecene relative. (m) Hypothetical intermediate

2.3. Sources

The genus *Fusarium* is classified according to the scheme proposed by Booth (*7*) and later extended and modified by Nelson *et al.* (*8, 9*), who, in the Sporotrichiella Section, accepted *F. chlamydosporium* but confined the name *F. tricinctum* to *F. tricinctum* sensu stricto. Nelson *et al.* (*9*) reexamined many of the known toxigenic *Fusarium* cultures and, in cases of mistaken identification, corrected the assignment. Where this has occurred the original assignment, or its equivalent in Booth's system of nomenclature, is placed in parentheses in Tables 1–4. Some new *Fusarium* sources claimed since this authoritative treatise (*9*) went to press (1981) have not been subjected to the same rigorous scrutiny.

F. compactum is equated with *F. equiseti* (*7*). The subspecies of *F. acuminatum*, *F. acuminatum* subsp. *armeniacum* (*234*) is not differentiated in the Tables. The morphological diversity of strains classified as *F. sambucinum* has led to their being divided into *F. sambucinum* sensu stricto, *F. torulosum* and *F. venenatum* (*235*). Strains classified as *F. torulosum* did not produce trichothecenes. Those classified as *F. sambucinum* sensu stricto produced mainly diacetoxyscirpenol, though some produced neosolaniol and T-2 toxin (*79, 236*), but *F. venenatum* strains produced only diacetoxyscirpenol (*79, 236*) and isotrichodermin (*20*) and their close relatives. In the Tables, "*F. sambucinum*" refers to *F. sambucinum* sensu lato. With the removal of *F. nivale* (Aarachnites Section) to the genus *Microdocium* (*237*), trichothecenes are produced by 20 *Fusarium* species drawn from 8 of the 12 Sections in Booth's scheme.

Stachybotrys atra, *S. alternans* and *S. chartarum* are synonymous (*238*) and *S. atra* is used. *Trichoderma lignorum* is equated with *T. viride* (*239*).

The phylogenetic relationship between *Spicellum roseum* and *Trichothecium roseum* has been examined by analysis of partial sequences of RNA subunits (*13*). Although the relationship was close, morphological differences supported the maintenance of separate genera for these sources. The genus *Memnoniella* is very close to *Stachybotrys* (*17*), but morphological differences are again sufficient for the two genera to be regarded as distinct.

Dendrodochium toxicum, a source of verrucarin A (*240*) and roridin A (*241*), is considered to be *Myrothecium verrucaria* (*242*).

The simple trichothecenes are products of *Acremonium*, *Fusarium*, *Myrothecium*, *Stachybotrys* and *Trichoderma* spp., and one species each of *Cephalosporium*, *Cylindrocladium*, *Dendrostilbella*, *Gliocladium*, *Memnoniella*, *Microdocium*, *Spicellum* and *Trichothecium*. *Trichothecium roseum* is, nevertheless, responsible for 12 metabolic products with the trichothecene skeleton. *Acremonium, Cylindrocladium, Gliocladium,* and *Memnoniella* are new sources reported since 1995.

The trichoverroids and fungal macrocycles are products of *Myrothecium* and *Stachybotrys* spp., and one species each of *Acremonium, Ceratopycnidium, Cercophora, Cylindrocarpon, Phoma, Phomopsis,* and *Verticinimonosporium. Acremonium, Ceratopycnidium, Cercophora* and *Phoma* are new sources reported since 1991.

The shrub *Baccharis artemisioides* is a new source of the known fungal macrocycles verrucarins A and J, and roridins A, D and E, as is the *Baccharis* spp. endophyte *Ceratopycnidium baccharidicola* (*164*). *B. megapotamica* and *B. coridifolia* remain the only sources of the baccharinoids.

Of the 98 known simple trichothecenes, 82 have been obtained from *Fusarium* spp. This is, in part, a reflection of the interest shown in toxins produced by the grain pathogens *F. sporotrichioides* (45 trichothecenes) and *F. graminearum* (33 trichothecenes).

Trichodiene (**49**; Scheme 1) has been detected as a volatile product of *Monascus purpureus* and of *Stachybotrys atra*, but the trichothecene relatives (Table 8), including, possibly, the loukacinols (see note c), are essentially products of *Fusarium* spp. and *T. roseum*.

2.4. Oxygenation Pattern

The oxygenation pattern of the simple trichothecenes shows marked genus specificity. With very few exceptions products of *Fusarium* spp. show oxygenation at C-3α, which can be accompanied by additional oxygenation at positions 15, 8α, 7α and 4β. On the other hand, products from *Myrothecium, Stachybotrys* and *Trichoderma* spp. show oxygenation at C-4β, which can be accompanied by additional oxygenation at positions 15 or 8α, and, in the case of *Myrothecium*, position 16. *Trichothecium roseum* holds an intermediate position with oxygenation at C-4β accompanied by additional oxygenation at C-8α and C-3α.

8β-Hydroxyscirpene (Table 1: $C_{15}H_{22}O_3$) (from *F. sporotrichioides*) is the only known example of 8β-hydroxylation in the simple trichothecenes.

12,13-Epoxytrichothec-9-en-8-ones (Table 2) are often referred to as Type B trichothecenes [Group III on the chemical classification (*2*), based on ring A chemistry]; the remaining simple trichothecenes, without a keto group at C-8, are Type A (Groups I and II of the chemical classification).

Within a given *Fusarium* species, variation in the ability to effect 4-hydroxylation or acetylation at positions 3 and 15 is significant, and the classification of some *Fusarium* strains in the Discolor Section

(*F. sambucinum, culmorum, graminearum, crookwellense*) into chemotypes based on the hydroxylation pattern of the metabolites has been proposed (*243*). Within *F. graminearum* two chemotypes exist, producing either vomitoxin (4-deoxynivalenol) or nivalenol-related trichothecenes (*244, 245, 246*). An important factor in all these chemotypes is the geographical location from which the organism was isolated. However, in assessing this work it must be remembered that some strains of *F. graminearum* produce both nivalenol and vomitoxin and/or their acetyl derivatives.

The sequences within the nuclear ribosomal DNA (28S) of a number of *Fusarium* spp. have been examined with a view to determining the genetic relationship of the trichothecene producers (*247*). It was concluded that the phylogenetic placement of these correlated better with secondary metabolite data than with the current classification system based on morphology. There seems to be general agreement that the systematics of the Fusaria is in an unsatisfactory state, and that, in the long term, a new system, based on genetic constitution, will be introduced.

The oxygenation pattern of the trichothecene nucleus in the fungal macrolides is similar to that of the simple trichothecenes, but roritoxin C is a $9\beta,10\beta$-epoxide and four of the five known products of an unidentified *Cylindrocarpon* sp. have a $7\beta,8\beta$-epoxide. A $7\beta,8\beta$-epoxide is also found in the simple trichothecene crotocin.

The baccharinoids (Table 7) are essentially roridins which have undergone further hydroxylation and/or epoxidation giving a substitution pattern found only rarely in the fungal macrocyclic trichothecenes. 3α-Hydroxylation, in baccharinoid B12, is unknown in other macrocycles, whilst 8β-hydroxylation occurs in ten baccharinoids, and $9\beta,10\beta$-epoxidation in six. 16-Hydroxylation occurs in three baccharinoids but otherwise only in non-macrocyclic *Myrothecium* products. Baccharinoid B27 is the only macrocycle to have an 8-oxo group, a common feature among simple trichothecenes.

3. Biosynthesis

3.1. Simple Trichothecenes

3.1.1. Mevalonic Acid to Trichodiene (Scheme 2)

The early work on the biosynthesis of the trichothecene nucleus related to the formation of trichothecolone (**101**) in *Trichothecium roseum* and of derivatives of verrucarol (**102**) in *Myrothecium* spp. This work has been reviewed in detail (*248*). The more recent work has been concerned mainly with the biosynthesis of T-2 toxin (**103**) by *Fusarium sporotrichioides* and of 3-acetylvomitoxin (**104**) by *F. culmorum* and *F. graminearum*.

Studies in which [2-^{13}C]-mevalonic acid lactone ("mevalonate") (**105**) was used as a precursor showed that three molecules were incorporated into trichothecolone (**101**), the label appearing at positions 4, 8, and 14 (*249*). The involvement of 2-*trans*-6-*trans*-farnesyldiphosphate (**106**) had been shown previously using ^{14}C-labelled material (*250*). The biosynthesis thus follows a standard sesquiterpene pathway from mevalonate through farnesyldiphosphate to nerolidyldiphosphate (**107**) (*251, 252*), which is then cyclised by the enzyme trichodiene synthase to the bicyclic hydrocarbon trichodiene (**49**) (*253*). The evidence for the participation of nerolidyldiphosphate has been reviewed (*254, 255*). None of the possible enzyme bound intermediates has been isolated from what is shown (**107, 108**) as a concerted process involving a 1,4-hydride shift and two 1,2-methyl shifts (*256*).

Results consistent with the same biosynthetic pathway to trichodiene have been obtained with *Fusarium graminearum* and *F. culmorum*, initially with labelled acetate (*257, 91*), and, conclusively, with [3,4-^{13}C$_2$]-mevalonate (*258*).

Trichodiene synthase has been obtained from *T. roseum* (*259*) and subsequently from *F. sporotrichioides* (*260, 261*) and *F. sambucinum* (*261*); the *F. sporotrichioides* enzyme has been characterised. Isomerisation of farnesyldiphosphate to nerolidyldiphosphate is the rate-determining step in the reaction pathway (*262*). Factors affecting substrate recognition by the active site have been studied (*263*), and a number of amino-acid

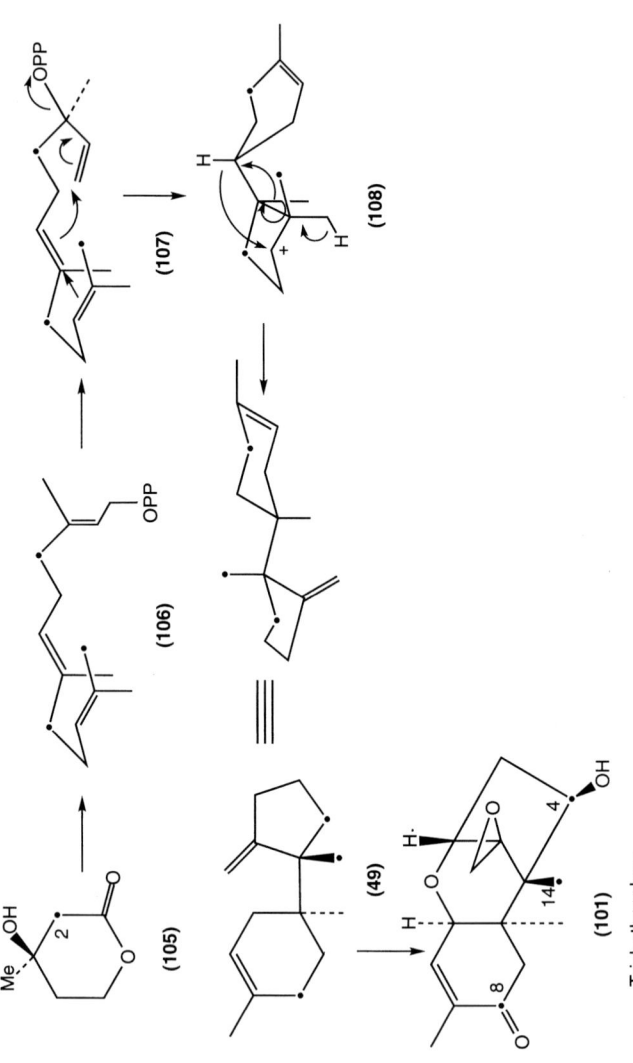

Scheme 2. Biosynthesis of trichodiene (49) and trichothecolone (101) from mevalonic acid lactone (105), labelled atom ●

residues important for the trichodiene synthase reaction have been identified (*264, 265*).

The trichodiene synthase gene, *Tri5*, has been cloned from *F. sporotrichioides* (*266*), *F. sambucinum* (*267*), *F. graminearum* (*268*), and *F. poae* (*269*); sequence analysis showed a high degree of similarity. The *F. sporotrichioides* gene has been expressed in *E. coli* (*270, 271*) and in tobacco, *Nicotiana tobaccum* (*272*). In *E. coli* the recombinant enzyme had properties closely resembling those of the native enzyme and acceptable yields of trichodiene were obtained: Only low yields of trichodiene were obtained in tobacco tissue. Using standard gene disruption techniques, $Tri5^-$ mutants have been obtained from plant pathogenic strains of *F. sambucinum* (*273, 267*) and *F. graminearum* (*268, 274*) in which trichothecene production is suppressed and the virulence of the strain is diminished. Wild-type virulence has been restored to *F. graminearum* $Tri5^-$ mutants by the reintroduction of *Tri5* (*274*). These results provide compelling evidence that trichothecene production contributes to the virulence of these strains.

Another important application arising from the work with trichodiene synthase has been the development of *Tri5* specific polymerase chain reaction-based assays for the detection of potential trichothecene-producing *Fusarium* species in pure culture and in contaminated grain (*269, 275, 276*).

3.1.2. Trichodiene to 12,13-Epoxytrichothecene and Isotrichodermol (Scheme 1)

Trichodiene (**49**), the last hydrocarbon intermediate in the pathway, has been isolated from *T. roseum*, from *Fusarium* spp., and also from *S. atra* (Table 8). Specific labelling experiments have shown it to be a precursor of trichodiol (**55**) and 12,13-epoxytrichothecene (**73**), in addition to trichothecolone (**101**), in *T. roseum* (*277*); it is also a precursor of isotrichodiol (**53**) (*219*) and 3-acetylvomitoxin (**104**) in *F. culmorum* (*278*). However, relatively little work has been done on this stage of the pathway to simple trichothecenes in *Myrothecium*, *Stachybotrys* and *Trichoderma* spp.

Much effort, using mainly *Fusarium* spp., has been devoted to the identification of the oxygenation, cyclisation and esterification steps which take place after the formation of trichodiene. Fermentations carried out on a large scale, or in the presence of enzyme inhibitors, such as ancymidol (*225*) or xanthotoxin (*219*), or with mutant strains (*223*), have yielded metabolic products [(**47**), (**52**), (**53**), (**59**): Table 8] which are proven intermediates. They are derived from trichodiene by plausible pathways involving allylic hydroxylation at positions 2α and 11α, followed or accompanied by β-epoxidation of the 12-ene. The number of potential intermediates points to the operation of metabolic grids rather than unique pathways (*279, 219*), but this aspect remains to be clarified.

A cell-free enzyme system capable of epoxidizing the 12-ene of a trichodiene derivative has been obtained from *F. culmorum* (*280*). This system was also capable of effecting 3α-hydroxylation of the product, an interesting result since 3α-hydroxylation is commonly found amongst the trichothecene relatives.

The *Tri4* gene of *F. sporotrichioides* encodes a cytochrome P450 enzyme which catalyses the first oxidation step post trichodiene (*281*). Although the product was not characterised, it was generally assumed to be 2α-hydroxytrichodiene (**47**). This compound has since been isolated from *F. culmorum* (*222*) together with the compound thought likely to be the second step in this stage of the pathway, isotrichool (**52**) (*225*). Specifically-labelled 2α-hydroxytrichodiene was incorporated into 3-acetylvomitoxin but synthetic 12,13-epoxytrichodiene was not, showing that 2α-hydroxylation precedes epoxidation. Specifically-labelled isotrichool was also a precursor of 3-acetylvomitoxin. The pathway function, if any, of 11α-hydroxytrichodiene (**48**) has not been determined.

Isotrichodiol (**53**) and isotrichotriol (**59**) and their respective products of acid-catalysed allylic rearrangement, trichodiol (**55**) and 9-epitrichodiol (**56**), both isolated from *T. roseum*, and trichotriol (**63**) and 9-epitrichotriol (**64**), isolated only from *Fusarium* spp., are central to the pathway from trichodiene to the epoxytrichothecene nucleus. The conversion of trichodiol and trichotriol into 12,13-epoxytrichothecene (**73**) and isotrichodermol (**74**), respectively, proceeds spontaneously, though slowly, at acid pH (*152*) by mechanism (A), Scheme 3. However, the rapid *in vivo* incorporation indicates an enzyme-mediated process (*282*). An alternative proposal, with mechanism (B), has isotrichodiol and isotrichotriol as the true intermediates (*231*).

There is ample circumstantial evidence that the next step in *Fusaria* in the pathway to T-2 toxin and 3-acetylvomitoxin after trichodiol/isotrichodiol is 3α-hydroxylation to trichotriol/isotrichotriol. In feeding studies with a mutant strain of *F. sporotrichioides* in which this step was blocked, trichotriol, 9-epitrichotriol and isotrichotriol were all converted into T-2 toxin, but trichodiol was not (*231*), indicating that 3α-hydroxylation precedes the cyclisation step leading to the trichothecene nucleus.

Arising from this important conclusion it was suggested (*283*) that the trichothecenes should be divided into two groups based on the structure of the pathway intermediate immediately preceding cyclization, d-type (*e.g. Myrothecium* and *Trichothecium* metabolites) derived from trichodiol; and t-type (*e.g.* most *Fusarium* metabolites) derived

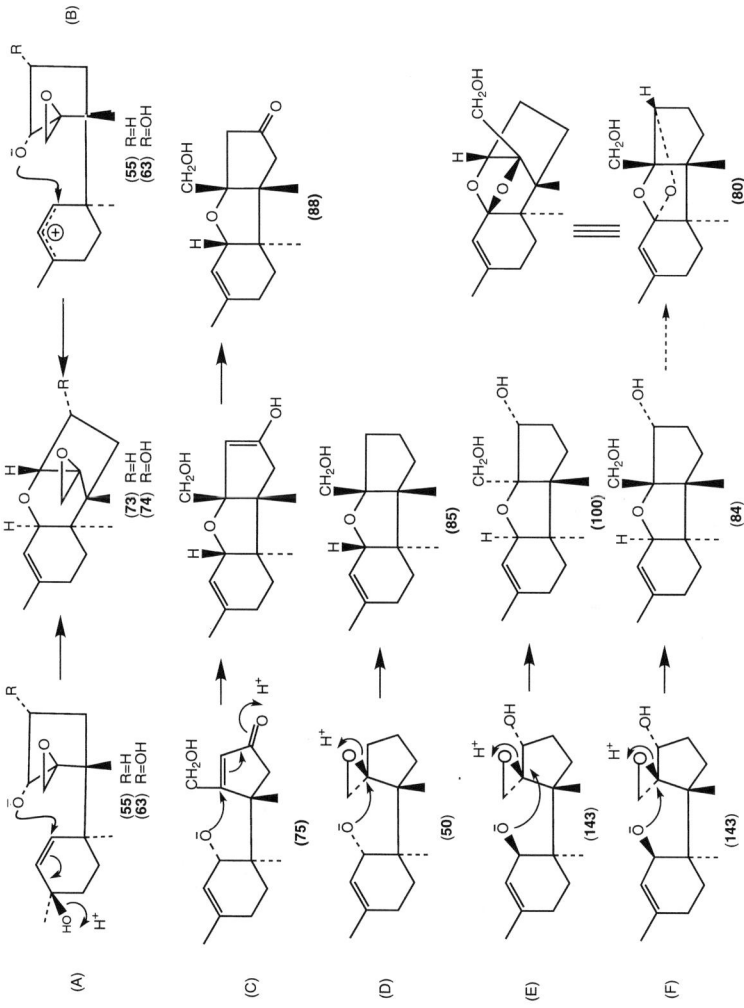

Scheme 3. Possible cyclisation mechanisms in the biosynthesis of trichothecenes and apotrichothecenes; the electronic representation is purely indicative, *e.g.* alcohol nucleophiles may act in their hydroxy rather than alkoxide forms

from trichotriol. Some doubt became attached to this attractive proposal, which was initially consistent with the known oxygenation pattern specificity, by the isolation from *T. roseum* of trichothecinols A (Table 2: $C_{19}H_{24}O_6$) and C (Table 1: $C_{19}H_{26}O_6$) which have 3α-hydroxylation.

There is no evidence that the 8- and 16-hydroxy substituted isotrichodiol and isotrichotriols isolated from *F. sporotrichioides* (Table 8) are, respectively, involved in the biosynthesis of 8- or 16-hydroxytrichothecenes.

3.1.3. Further Oxygenation and Esterification of the Trichothecene Nucleus (Scheme 4): Biosynthesis of Specific Metabolites

3.1.3.1. Trichothecolone

In *T. roseum* fermentations, experiments with specifically-labelled materials showed that both trichodermol (**110**) and trichothecodiol (**112**) were converted into trichothecolone (**101**) and thence to trichothecin (**113**) (*284*), indicating that 4-hydroxylation of 12,13-epoxytrichothecene (**109**) probably precedes 8-hydroxylation in the biosynthetic sequence: (**109**) → (**110**) → (**112**) → (**101**). The *T. roseum* metabolite (**109**) has always been assumed (*10*), but, to this reviewer's knowledge, never proved, to be a trichothecolone precursor.

3.1.3.2. Vomitoxin and Derivatives

Although isotrichodermol (**114**) [or its acetate isotrichodermin (**115**)] was found to be a precursor both of 3-acetylvomitoxin (**104**) in *F. culmorum* (*226*), and of T-2 toxin (**103**) using the blocked mutant strain (see above) of *F. sporotrichioides* (*232*), 12,13-epoxytrichothecene (**109**) was not (*285*)[3] (*232*). 12,13-Epoxytrichothecene was not converted into isotrichodermol by *F. culmorum* (*285*), and neither trichodermol (**110**) nor 15-hydroxy-12,13-epoxytrichothecene were precursors of T-2 toxin in *F. sporotrichioides* (*232*). These results reinforce the earlier conclusion (above) that 3α-hydroxylation of trichodiol precedes the cyclization step in these organisms. The conversion of isotrichodermol to isotrichodermin by an acetyltransferase is discussed in connection with the function of the resistance gene *Tri101* (Section 3.1.4.).

The steps in the conversion of isotrichodermin (**115**) to 3-acetylvomitoxin (**104**) follow logically from the pattern of secondary metabolites, common to *F. culmorum* and *F. graminearum*, but originally

---

[3] Correcting an earlier (*226*) statement to the contrary.

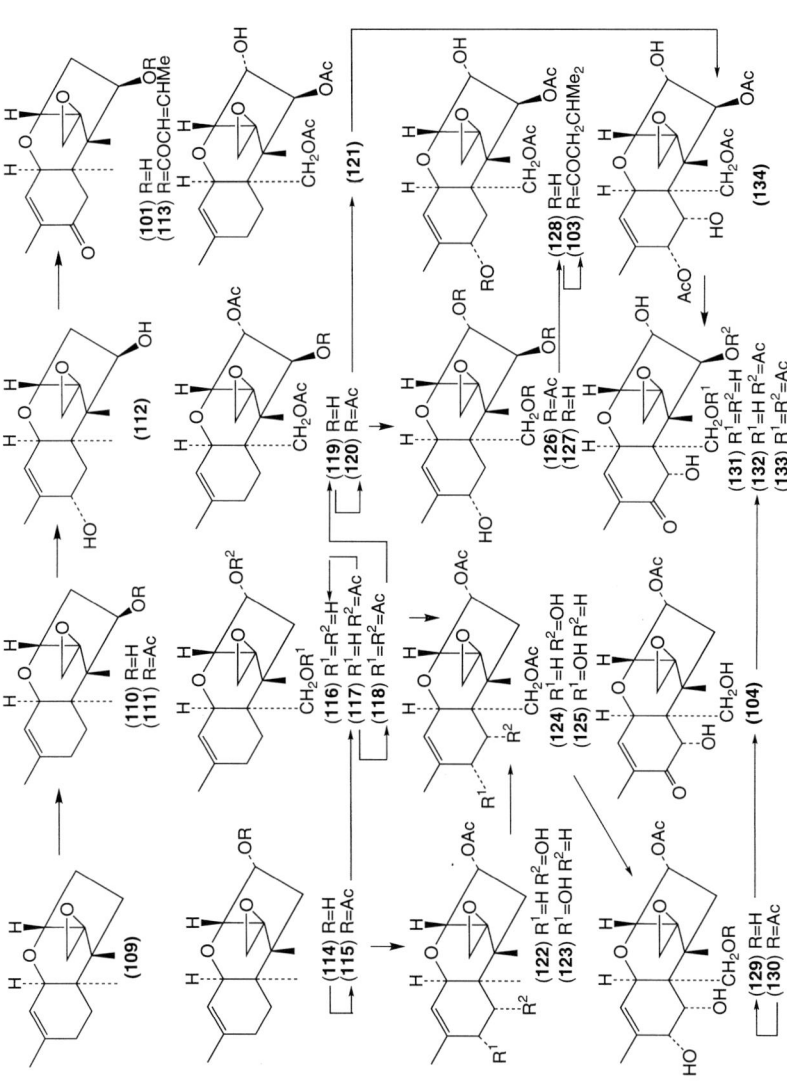

Scheme 4. Biosynthetic pathway to trichothecolone (**101**) from 12,13-epoxytrichothecene (**109**), and to 4,15-diacetoxyscirpenol (**121**), T-2 toxin (**103**), and 3-acetylvomitoxin (**104**) from isotrichodermol (**114**)

isolated from a *F. graminearum* fermentation (*42*). However, this pattern, consisting of 7-(**122**), 8-(**123**), and 15-hydroxyisotrichodermin (15-deacetylcalonectrin) (**117**), 7-(**124**), and 8-hydroxycalonectrin (**125**) and 7,8-dihydroxycalonectrin (**130**) gave no clue as to the sequence in which hydroxylation occurred. Kinetic pulse-labelling techniques identified 15-deacetylcalonectrin (**117**) as the first intermediate after isotrichodermin in *F. culmorum* (*44*). This step has been confirmed with a cell-free enzyme system (*286*); however, this system was not regiospecific and also hydroxylated isotrichodermin at positions 7 and 8.

7-(**122**), 8-(**123**), and 15-Hydroxyisotrichodermin (**117**), and calonectrin (**118**) are all proven precursors of 3-acetylvomitoxin (**104**) in *F. culmorum* (*44*), consistent with the operation of a metabolic grid. The 7,8-dihydroxy compounds (**129**) and (**130**), derivatives, respectively, of 15-deacetylcalonectrin and calonectrin, have both been shown to be precursors of 3-acetylvomitoxin, but the deacetylcalonectrin derivative (**129**) was not incorporated into dihydroxycalonectrin (**130**) (*287*). This result indicates a unique pathway for the last two steps in 3-acetylvomitoxin biosynthesis in *F. culmorum*, namely 15-deacetylation of dihydroxycalonectrin (**130**) followed by oxidation at C-8.

There is, nevertheless, evidence that this section of the pathway from isotrichodermin (**115**) to 3-acetylvomitoxin (**104**) is not as straightforward as depicted in Scheme 4, and, overall, at least one deacetylation/reacetylation step takes place at position 3 with the involvement of isoverrucarol ("dideacetylcalonectrin") (**116**) (*288*) in the pathway. Esterases with a high specificity for the 3-position have been obtained from an unidentified *Fusarium* sp. (*289*) and from *F. sporotrichioides* (see below).

The emphasis of most of this work has been on 3-acetylvomitoxin biosynthesis. Both 15-acetylvomitoxin and vomitoxin are major metabolic products of some strains of *F. graminearum* (*147*), but the steps involved in the late stages of their biosynthesis have not yet been examined.

The *Tri11* gene of *Fusarium sporotrichioides* encodes a cytochrome P450 hydroxylase which converts isotrichodermin to 15-deacetylcalonectrin (*290*). Although the *Tri11* gene from *F. graminearum* has substantial identity with the *F. sporotrichioides* gene, and presumably encodes a P450 monooxygenase (*291*), the *F. culmorum* enzyme is not a hemoprotein and is not attached to P450 (*291*). *Tri11*$^-$ mutant strains, lacking a functional C-15 hydroxylase, accumulate the 8α-hydroxy-derivatives (**135**; R = H and OH) of isotrichodermol and scirpene-3,4-diol, respectively (*23*).

The genes and enzymes involved in the steps between calonectrin and the vomitoxin derivatives, particularly the final oxidation step which differentiates Type B from Type A compounds, have still to be identified.

References, pp. 113–130

(135)

3.1.3.3. T-2 Toxin

The *Tri3* gene of *F. sporotrichioides* encodes a 15-O-acetyl-transferase that acetylates the 15-OH of 15-deacetylcalonectrin (**117**) (*292*). This enzyme also acetylates the 3α-hydroxyl in a number of 12,13-epoxytrichothecene substrates. The C-15 acetylation step appears to be essential to the pathway to T-2 toxin and *Tri3*⁻ mutants accumulate neither T-2 toxin nor the major co-metabolites 4,15-diacetoxyscirpenol (**121**) and neosolaniol (**128**).

It seems certain that the pathways to 3-acetylvomitoxin and T-2 toxin in *Fusaria* diverge at 15-deacetylcalonectrin; the next step on the T-2 toxin pathway, after 15-acetylation, is the 4β-hydroxylation of calonectrin to 3,15-diacetoxyscirpenol (**119**). Nothing is known about the genes and enzymes involved with C-4 hydroxylation in *Fusaria*. The gene *Tri7* controls the 4-OH acetylation step in *F. sporotrichioides* (*293*), giving triacetoxy-scirpene (**120**). The gene disruption mutant *Tri7*⁻ did not produce T-2 toxin, but accumulated HT-2 toxin (4-deacetylT-2 toxin); and feeding studies with this mutant showed that only precursors containing a 4-acetyl group, *e.g.* 4,15-diacetoxyscirpenol (**121**), were converted to T-2 toxin, whilst those without this grouping, *e.g.* HT-2 toxin, were not.

C-4 acetylation is followed by 8α-hydroxylation to 3-acetylneosolaniol (**126**) and esterification at C-8. The *Tri1* gene controls C-8 hydroxylation in *F. sambucinum* (*294*), and *Tri8* is said to have a similar function in *F. sporotrichioides* (*295, 293*). However, there is evidence that *Tri8* is involved in both oxidation and esterification steps at C-8. A *Tri8*⁻ mutant strain did not produce T-2 toxin but accumulated 4,15-diacetoxyscirpenol, suggesting that *Tri8* is involved specifically in C-8 oxidation. Additionally though, this mutant was unable to effect esterification with the isovalerate moiety, and failed to convert T-2 tetraol (**127**) to T-2 toxin. A known T-2 toxin-deficient *F. sporotrichioides* strain, produced by UV mutagenesis, was able to effect this esterification.

4,15-Diacetoxyscirpenol (**121**), neosolaniol (**128**) and T-2 toxin (**103**) result from deacetylation of their respective 3-acetyl derivatives.

An esterase with higher affinity for the ester bond at the 3α position than at the 4β position and with some specificity for 3-acetylT-2 toxin has been partially purified and characterised from *F. sporotrichioides* (*296*).

The isovalerate moiety present in T-2 toxin is derived by decarboxylative transamination from L-leucine (*297, 298*). Leucine limitation enhances the production of neosolaniol (**128**).

3.1.3.4. Nivalenol and Derivatives

By using a *Fusarium* strain known to be an abundant producer of diacetoxyscirpenol, 3-acetylvomitoxin (**104**) was converted to fusarenone (**132**) (*299*).

The sequence of hydroxylation steps leading to nivalenol (**130**) and its derivatives is unproven, but could follow logically from the pattern of secondary metabolites isolated from *F. equiseti* (*5*): 4,15-diacetoxyscirpenol (**121**) → the diol (**134**) → 4,15-diacetylnivalenol (**133**). The genes and enzymes concerned with these (hypothetical) oxidative steps have not been investigated. The diol (**134**) has not been reported from any other *Fusarium* spp.

The work outlined in Sections 3.1.3.2. and 3.1.3.3., above, deals mainly with the biosynthesis of 3-acetylvomitoxin by *F. culmorum* and of T-2 toxin by *F. sporotrichioides*; some strains of both these *Fusaria* produce nivalenol and derivatives, possibly by appropriate modification of the vomitoxin and T-2 toxin pathways.

3.1.4. Trichothecene Biosynthetic Gene Clusters

In *F. sporotrichioides* ten biosynthetic pathway genes, *Tri3–12*, are closely linked and form a gene cluster (*300*). Homologous gene clusters exist in *F. sambucinum* (*301*) and in *F. graminearum* (*302, 293*). In an important new approach to the investigation of trichothecene diversity, the biosynthetic gene clusters from T-2 toxin- and vomitoxin-producing strains of *F. sporotrichioides* and *F. graminearum*, respectively, were subjected to comparative analysis of nucleotide sequence and genome organization and orientation (*293*). The *Tri7* gene from *F. graminearum* was found to include base deletions and insertions which would preclude translation of a functional protein, consistent with the absence of C-4 oxygenated products from this strain. Similar comparative analyses of Type A- and Type B-producing strains of *F. sporotrichioides*, and of vomitoxin- and nivalenol-producing chemotypes of *F. graminearum*, could yield interesting results.

References, pp. 113–130

The genes *Tri1* from *F. sambucinum* (*294*) and *Tri101* from *F. graminearum* (*283*) are not linked to the clusters. In addition to the six genes, *Tri3–5,7,8* and *11*, that control biosynthetic enzymes, whose function has been outlined above, the *F. sporotrichioides* and *F. graminearum* clusters contain one regulatory gene, *Tri6*, and one membrane facilitator gene, *Tri12*. *Tri9* and *Tri10* encode proteins whose function is not understood (*293*).

Tri6 encodes a zinc finger protein involved in the transcriptional regulation of the pathway genes concerned with trichothecene biosynthesis and binds with all but one of the known pathway genes (*303*). Disruption of *Tri6* resulted in a mutant that accumulated low levels of trichodiene but did not produce trichothecenes, was unable to convert the intermediates calonectrin, diacetoxyscirpenol and neosolaniol to T-2 toxin, and had greatly reduced transcription of *Tri4* (*304*). The *Tri6* homologs cloned from the Type B trichothecene producers *F. graminearum* and *F. crookwellense* were almost identical at the amino acid sequence level to *Tri6* from *F. sporotrichioides*, a Type A producer (*305*).

The *Tri12* gene encodes a protein involved in the specific extrusion of trichothecenes across the fungal plasma membrane, causing their accumulation in the substrate and thus participating in the self-defence mechanism of the producing organism (*306*). *Tri102* from *F. graminearum* (*302*) is identical with *Tri12*.

Tri101, a pathway gene not associated with the gene cluster, is also involved with resistance of the producing organism to toxic metabolites. It encodes an 3α-acetyltransferase, an esterification which significantly diminishes the toxicity, including phytotoxicity (*307*), of a trichothecene, and has been cloned from both *F. graminearum* (*283*, *308*) and *F. sporotrichioides* (*309*). More recent work with *Tri101*⁻ mutants has shown that *Tri101* converts isotrichodermol to isotrichodermin and is required for the biosynthesis of T-2 toxin by *F. sporotrichioides* (*310*). The results are consistent with the hypothesis that much of the pathway to T-2 toxin involves 3-acetylated intermediates (*283*, *310*).

3.2. Trichoverroids and Macrocyclic Trichothecenes (Scheme 5)

The genes controlling the initial stages of the biosynthetic pathway to the verrucarol moiety of the macrocyclic trichothecenes have been studied in *Myrothecium roridum* (*311*). In so far as the pathways overlap, the pathway outlined above for simple trichothecenes is followed, and clustered genes *MRTri4–6*, similar to *Tri4–6* and with the same function,

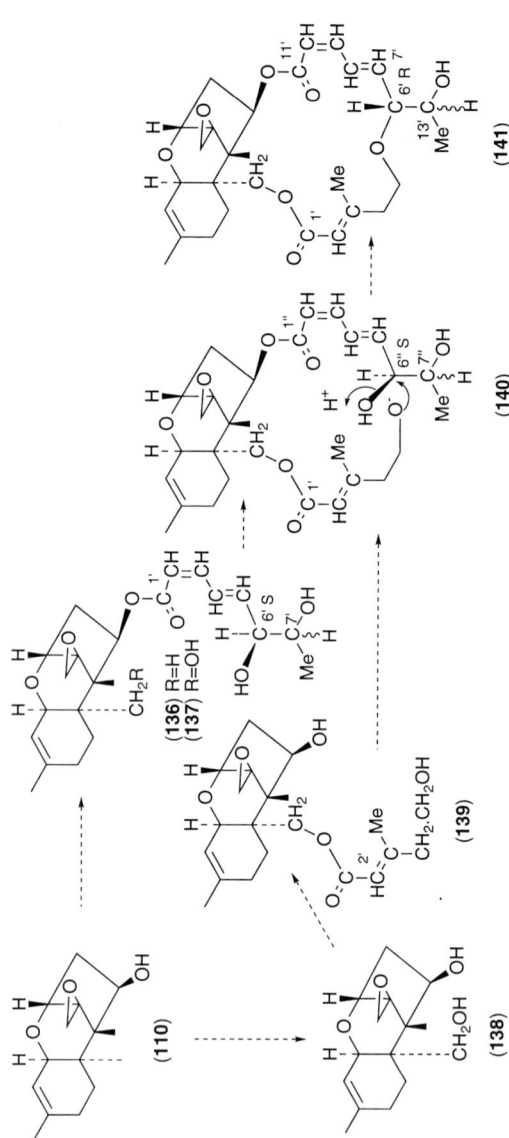

Scheme 5. Possible pathways from trichodermol (**110**) to roridin E (**141**, 13′-βH)

have been cloned. There are differences, however, and the protein specified by *MRTri6* is almost twice the size of the analog from *F. sporotrichioides*.

The sequence of the hydroxylation and esterification steps in *Myrothecium* spp. fermentations between trichodermol (**110**) and the C-7′ diastereoisomeric trichoverrins (C-6″S) (**140**) and isotrichoverrins (C-6″R) is unclear and has still to be tested with specifically labelled compounds. The pattern of secondary metabolites which could be intermediates, verrucarol (**138**), verrol (**139**), the trichodermadienediols (**136**) and the trichoverrols (**137**), again suggests a metabolic grid rather than a unique pathway.

The six-carbon moiety attached to C-15 in verrol and the trichoverrins is mevalonate-derived (*248*), and after cyclisation to form the macrolide, the 2′-ene can undergo hydrogenation, α-epoxidation, or hydration. The ethylenic double bonds in the acetate-polymalonate-derived eight-carbon moiety attached to C-4 in the trichoverroids occur as Z or E (see Table 1), but are always 7′ (E), 9′(Z) in the macrolide.

Ring closure of the trichoverrins occurs with inversion of configuration at position 6′. In most roridins and all baccharinoids C-6′ is R; the corresponding centre in the trichoverrins is S. Exceptionally, iso-roridin E, 13′-epiiso-roridin E and 6′-epi-13′-epiroridin A have C-6′S, but the *M. verrucaria* strains which produce them also produce iso trichoverrins with C-6′R (*113*). Trichoverrin B has been proved to be a precursor of roridin A (**28**; $R^1 = R^4 = H$, $R^2 = R^3 = \beta H$) and verrucarin A (**7**; $R^1 = H$, $R^2 = OH$) (*105*).

It is likely that the trichoverroid roridin L-2 (**142**) and its close relatives arise by hydrolysis and relactonisation of the corresponding, unknown, 12′-hydroxyroridin E analogues.

(**142**)

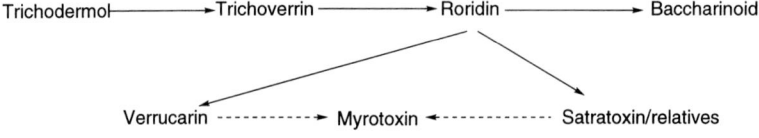

Scheme 6. Possible biosynthetic pathway to the macrocyclic trichothecenes

C-7′-diastereoisomeric 12,13-deoxytrichoverrins have been isolated from *M. verrucaria* but whether these 12-enes are precursors of verrucarin K, or whether deoxygenation takes place after the formation of the macrolide is unknown. Likewise, a possible relationship between 8α-hydroxyisotrichoverrin A (Table 1: $C_{29}H_{40}O_{10}$) and verrucarin L is unproven.

A plausible sequence based on inspection of the structural formulae, and in which the roridin skeleton holds a central position, is shown in Scheme 6. The scheme assumes that the pathway in *Baccharis*, or its endophyte (see below), is essentially the same as in *Myrothecium*, and that all the fungal producers of macrocyclic trichothecenes use the same pathway.

The biosynthesis of the macrocyclic trichothecenes in *Baccharis* remains controversial. It was initially believed (*188*) that the baccharinoids were the products of a plant-fungus interaction in which fungal macrocycles were taken up by the plant and further modified. This hypothesis was withdrawn (*162*) following the failure to demonstrate any fungal involvement either from the immediate soil environment or from examination by electron microscopy of the plant tissues. It was concluded (*162, 312*) that *Baccharis* synthesized macrocyclic trichothecenes *de novo*. Some doubts about this conclusion have resulted as a result of the isolation (*313*) from *B. coridifolia* of an endophyte, *Ceratopycnidium baccharidicola*, and the demonstration (*164, 314*) that this organism is a source of verrucarins and roridins (see Tables 5 and 6), all of which are known metabolic products attributed to *B. coridifolia*. However, none are baccharinoids, suggesting that the original plant-fungus interaction hypothesis (*188*) could have been correct.

3.3. Trichothecene Relatives (Scheme 1)

This account deals only with the 11-epimeric apotrichothecenes and with sambucinol (**81**). A more extended account covering all the trichothecene relatives derived from trichodiene is contained in Ref. 3.

Of the naturally occurring apotrichothecenes, compounds in the 11-epi-series are the most numerous, but little work has been done on their biosynthesis. The *F. sambucinum* metabolites FS 1 (**75**) and FS 4 (**67**),

References, pp. 113–130

which bear the same relationship to each other as do isotrichodiol and trichodiol, are possible precursors of dehydroapotrichodiol (**88**) and hence of the 3-epimeric apotrichodiols (**86**) and (**87**) by mechanism C, Scheme 3.

Alternatively, by mechanism D, the hypothetical intermediate (**50**) could lead to apotrichool (**85**) directly. Apotrichool was converted by *F. culmorum* into apotrichodiol (**86**) (*207*).

The apotrichothecene (**83**) is the product from the acidcatalysed trichothecene-apotrichothecene rearrangement of 12,13-epoxytrichothecene (**73**) and could be an artefact. The 12-epi-apotrichothecene (**100**) is believed to arise from the hypothetical intermediate 11-epiisotrichodiol (**143**: Scheme 3), and by mechanism E (*228*) which involves epoxide-opening with retention of configuration. Isotrichodiol and its 11-epimer should be interconvertible via the common allylic cation.

The nature of the biosynthetic pathway to sambucinol (**81**) is controversial. Trichodiene and isotrichool (**52**) are proven precursors in *F. culmorum* (*225*), as are apotrichool (**85**) (*226*) and 3-deoxysambucinol (**80**) (*285*). Another precursor is $2\alpha,13$-dihydroxyapotrichothecene (**84**) (*227*), which has not yet been isolated as a natural product but which should be readily obtainable from 11-epiisotrichodiol (**143**) by mechanism F, Scheme 3. Apotrichodiol (**86**) is converted photochemically in solution at room temperature into the acetal (**92**) (*211*), a likely precursor of sporol (**91**). Similar photochemical reactions with 2α-hydroxyapotrichothecenes, which could lead to deoxysambucinol (**80**) (mechanism F), have not been reported.

12,13-Epoxytrichothecene (**73**) has also been identified as a precursor of sambucinol (*226*), but the evidence for this has been criticised (*211*). There is no evidence for the participation of the 11-ketone (**72**) in the pathway from isotrichool to sambucinol (*227*).

References

1. Grove JF (1993) Macrocyclic Trichothecenes. Nat Prod Rep **10**: 429
2. Grove JF (1988) Non-Macrocyclic Trichothecenes. Nat Prod Rep **5**: 187
3. Grove JF (1996) Non-Macrocyclic Trichothecenes, Part 2. Prog Chem Org Nat Prod **69**: 1
4. Godtfredsen WO, Grove JF, Tamm Ch (1967) Zur Nomenklatur einer neueren Klasse von Sesquiterpenen. Helv Chim Acta **50**: 1666
5. Brian PW, Dawkins AW, Grove JF, Hemming HG, Lowe D, Norris GLF (1961) Phytotoxic Compounds Produced by *Fusarium equiseti*. J Exp Bot **12**: 1
6. Jarvis BB, Stahly GP, Pavanasasivam G, Midiwo JO, DeSilva T, Holmlund CE, Mazzola EP, Geoghegan RF (1982) Isolation and Characterization of the Trichoverroids and New Roridins and Verrucarins. J Org Chem **47**: 1117
7. Booth C (1971) The Genus *Fusarium*. Commonwealth Mycological Institute, Kew, UK

8. Nelson PE, Toussoun TA, Marasas WFO (1983) *Fusarium* Species: An Illustrated Manual for Identification. Pennsylvania State University Press, University Park, PA
9. Marasas WFO, Nelson PE, Toussoun TA (1984) Toxigenic *Fusarium* Species. Identity and Mycotoxicology. Pennsylvania State University Press, University Park, PA
9a. Kurata H (1984) Personal communication, in ref. 9, p. 128
10. Machida Y, Nozoe S (1972) Biosynthesis of Trichothecin and Related Compounds. Tetrahedron **28**: 5113
11. Baldwin NCP, Bycroft BW, Dewick PM, Marsh DC, Gilbert J (1987) Trichothecene Mycotoxins from *Fusarium culmorum* Cultures. Z Naturforsch **C42**: 1043
12. Lauren DR, Sayer ST, di Menna ME (1992) Trichothecene Production by *Fusarium* Species Isolated from Grain and Pasture Throughout New Zealand. Mycopathologia **120**: 167
13. Seifert KA, Louis-Seize G, Savard ME (1997) The Phylogenetic Relationships of Two Trichothecene-Producing Hyphomycetes, *Spicellum roseum* and *Trichothecium roseum*. Mycologia **89**: 250
14. Harri E, Loeffler W, Sigg HP, Stahelin H, Stoll Ch, Tamm Ch, Wiesinger D (1962) Über die Verrucarine und Roridine, eine Gruppe von cytostatisch hochwirksamen Antibiotica aus *Myrothecium*-Arten. Helv Chim Acta **45**: 839
15. Adams PM, Hanson JR (1972) Sesquiterpenoid Metabolites of *Trichoderma polysporum* and *T. sporulosum*. Phytochemistry **11**: 423
16. Ayer WA, Miao S (1993) Secondary Metabolites of the Aspen Fungus *Stachybotrys cylindrospora*. Canad J Chem **71**: 487
17. Jarvis BB, Zhou Y, Jiang J, Wang S, Sorenson WG, Hintikka E-L, Nikulin M, Parikka P, Etzel RA, Dearborn DG (1996) Toxigenic Molds in Water-Damaged Buildings: Dechlorogriseofulvins from *Memnoniella echinata*. J Nat Prod **59**: 553
18. Choi SU, Choi EJ, Kim KH, Kim NY, Kwon B-M, Kim SU, Bok SH, Lee SY, Lee CO (1996) Cytotoxicity of Trichothecenes to Human Solid Tumor Cells *in Vitro*. Arch Pharm Res **19**: 6
19. Lauren DR, Ashley A, Blackwell BA, Greenhalgh R, Miller JD, Neish GA (1987) Trichothecenes Produced by *Fusarium crookwellense* DAOM 193611. J Agric Food Chem **35**: 884
20. Miller JD, MacKenzie S (2000) Secondary Metabolites of *Fusarium venenatum* Strains with Deletions in the *Tri5* Gene Encoding Trichodiene Synthase. Mycologia **92**: 764
21. Corley DG, Rottinghaus GE, Tracy JK, Tempesta MS (1986) New Trichothecene Mycotoxins of *Fusarium sporotrichioides* (MC-72083). Tetrahedron Lett **27**: 4133
22. Pineiro MS, Scott PM, Kanhere SR (1996) Mycotoxin Producing Potential of *Fusarium graminearum* Isolates from Uruguayan Barley. Mycopathologia **132**: 167
23. McCormick SP, Hohn TM (1997) Accumulation of Trichothecenes in Liquid Cultures of a *Fusarium sporotrichioides* Mutant Lacking a Functional Trichothecene C-15 Hydroxylase. Appl Environ Microbiol **63**: 1685
24. Jarvis BB, Lee Y-W, Yatawara CS, Mazzocchi DB, Flippen-Anderson JL, Gilardi R, George C (1985) 7α-Hydroxytrichodermol, a New Trichothecene from *Myrothecium roridum*. Appl Environ Microbiol **50**: 1225
25. El-Kady IA, Moubasher MH (1982) Toxigenicity and Toxins of *Stachybotrys* Isolates from Wheat Straw Samples in Egypt. Exptl Mycol **6**: 25
26. Ueno Y, Aikawa Y, Okumura H, Sugura Y, Nakamura K, Masuma R, Tanaka T, Young CJ, Savard ME (1997) Trichothecenes Produced by *Fusarium* Species Fn 2B. Maikotokishin **45**: 25; (1998) Chem Abstr **128**: 44831

27. Greenhalgh R, Levandier D, Adams W, Miller JD, Blackwell BA, McAlees AJ, Taylor A (1986) Production and Characterization of Deoxynivalenol and Other Secondary Metabolites of *Fusarium culmorum* (CMI 14764, HLX 1503). J Agric Food Chem **34**: 98
28. Plattner RD, Tjarks LW, Beremand MN (1989) Trichothecenes Accumulated in Liquid Culture of a Mutant of *Fusarium sporotrichioides* NRRL 3299. Appl Environ Microbiol **55**: 2190
29. Kim K-H, Lee Y-W, Mirocha CJ, Pawlosky RJ (1990) Isoverrucarol Production by *Fusarium oxysporum* CJS-12 Isolated from Corn. Appl Environ Microbiol **56**: 260
30. Pathre SV, Mirocha CJ, Christensen CM, Behrens J (1976) Monoacetoxyscirpenol. A New Mycotoxin Produced by *Fusarium roseum* Gibbosum. Appl Environ Microbiol **24**: 97
31. Sobolev VS, Eller KI, Boltyenskaya EV, Dmitrieva IV, Tutelyan VA (1984) Study of Toxin Production by *Fusarium sporotrichiella*. Isv Akad Nauk SSSR, Ser Biol 137; Chem Abstr **100**: 171231
32. Sanson DR, Corley DG, Barnes CL, Searles S, Schlemper EO, Tempesta MS, Rottinghaus GE (1989) New Mycotoxins from *Fusarium sambucinum*. J Org Chem **54**: 4313
33. Luo Y, Lin F, Yang J, Zhang J, Ye Y, Li Y, Jiang Y, Zhang N, Wang Z (1993) Analysis and Identification of *Fusarium* Mycotoxins in Corn Culture of *Fusarium camptoceras*. Zhenjun Xuebao **12**: 77; Chem Abstr **119**: 153704
34. McLachlan A, Shaw KJ, Hocking AD, Pitt JI, Nguyen THL (1992) Production of Trichothecene Mycotoxins by Australian *Fusarium* Species. Food Addit Contam **9**: 631
35. Liu W, Sundheim L, Langseth W (1998) Trichothecene Production and the Relationship to Vegetative Compatibility Groups in *Fusarium poae*. Mycopathologia **140**: 105
36. O'Donnell K, Cigelnik E, Casper HH (1998) Molecular Phylogenetic, Morphological, and Mycotoxin Data Support Reidentification of the Quorn Mycoprotein Fungus as *Fusarium venenatum*. Fungal Genet Biol **23**: 57
37. Szathmary CI, Mirocha CJ, Palyusik M, Pathre SV (1976) Identification of Mycotoxins Produced by Species of *Fusarium* and *Stachybotrys* Obtained from Eastern Europe. Appl Environ Microbiol **32**: 579
38. Cole RJ, Dorner JW, Cox RH, Cunfer BM, Cutler HG, Stuart BP (1981) The Isolation and Identification of Several Trichothecene Mycotoxins from *Fusarium heterosporum*. J Nat Prod **44**: 324
39. Kononenko GP, Soboleva NA, Leonov AN (1990) 3,7,8,15-Tetrahydroxy-12,13-epoxytrichothec-9-ene in a Culture of *Fusarium graminearum*. Khim Prir Soedin **267**; Chem Nat Compd (Engl Transl) **26**: 219
40. Godtfredson WO, Vangedal S (1965) Trichodermin, a New Sesquiterpene Antibiotic. Acta Chem Scand **19**: 1088
41. Cutler HG, Cole PD, Arrendale RF, Bassfield RL, Cox RH, Roberts RG, Hanlin RT (1986) *Dendrostilbella* sp. (85–25), a New Source of Trichodermin. Agric Biol Chem **50**: 2667
42. Greenhalgh R, Meier R-M, Blackwell BA, Miller JD, Taylor A, ApSimon JW (1984) Minor Metabolites of *Fusarium roseum* (ATCC 28114). J Agric Food Chem **32**: 1261
43. Greenhalgh R, Meier R-M, Blackwell BA, Miller JD, Taylor A, ApSimon JW (1986) Minor Metabolites of *Fusarium roseum* (ATCC 28114), Part 2. J Agric Food Chem **34**: 115
44. Zamir LO, Devor KA, Sauriol F (1991) Biosynthesis of the Trichothecene 3-Acetyldeoxynivalenol. Identification of the Oxygenation Steps After Isotrichodermin. J Biol Chem **266**: 14992

45. Gardner D, Glen AT, Turner WB (1972) Calonectrin and 15-Deacetylcalonectrin, New Trichothecanes from *Calonectria nivalis*. J Chem Soc, Perkin Trans 1: 2576
46. Loeffler W, Mauli R, Ruesch ME, Stahelin H (1967) Monodeacetylanguidin. German Patent 1233098, Jan 26, 1967; Chem Abstr **66**: 84744
47. Steyn PS, Vleggaar R, Rabie CJ, Kriek NPJ, Harington JS (1978) Trichothecene Mycotoxins from *Fusarium sulphureum*. Phytochemistry **17**: 949
48. Bergers WWA, van der Stap JGMM, Kientz CE (1985) Trichothecene Production in Liquid Stationary Cultures of *Fusarium tricinctum* NRRL 3299 (Synonym: *F. sporotrichioides*): Comparison of Quantitative Brine Shrimp Assay with Physicochemical Analysis. Appl Environ Microbiol **50**: 656
49. Evidente A, Randazzo G, Visconti A, Bottalico A (1989) Isolation of 15-Acetoxyscirpendiol from a Culture of *Fusarium poae* on Corn. Mycotoxin Res **5**: 30
50. Ichihara A, Katayama K, Teshima H, Oikawa H, Sakamura S (1996) Chaetoglobosin O and Other Phytotoxic Metabolites from *Cylindrocladium floridanum*, a Causal Fungus of Alfalfa Black Rot Disease. Biosci Biotech Biochem **60**: 360
51. Ishii K, Pathre SV, Mirocha CJ (1978) Two New Trichothecenes Produced by *Fusarium roseum*. J Agric Food Chem **26**: 649
52. Richardson KE, Toney GE, Haney CA, Hamilton PB (1989) Occurrence of Scirpentriol and Its Seven Acetylated Derivatives in Culture Extracts of *Fusarium sambucinum* NRRL 13495. J Food Prot **52**: 871
53. Visconti A, Mirocha CJ, Logrieco A, Bottalico A, Solfrizzo M (1989) Mycotoxins Produced by *Fusarium acuminatum*. Isolation and Characterization of Acuminatin: A New Trichothecene. J Agric Food Chem **37**: 1348
54. Corley DG, Rottinghaus GE, Tempesta MS (1986) Novel Trichothecenes from *Fusarium sporotrichioides*. Tetrahedron Lett **27**: 427
55. Hesketh AR (1992) Metabolic Studies on the Transformation of Trichodiene to Trichothecene Mycotoxins. Mycotoxin Res **8**: 52
56. Ishii K, Ueno Y (1981) Isolation and Characterization of Two New Trichothecenes from *Fusarium sporotrichioides* Strain M-1-1. Appl Environ Microbiol **42**: 541
57. Glaz ET, Csanyi E, Gyimesi J (1966) Supplementary Data on Crotocin – An Antifungal Antibiotic. Nature (London) **212**: 617
58. Achilladelis B, Hanson JR (1969) Minor Terpenoids of *Trichothecium roseum*. Phytochemistry **8**: 765
59. Plattner RD, Al-Hetti MB, Weisleder D, Sinclair JB (1988) A New Trichothecene from *Trichothecium roseum*. J Chem Res (S): 311
60. Langley P, Shuttleworth A, Sidebottom PJ, Wrigley SK, Fisher PJ (1990) A Trichothecene from *Spicellum roseum*. Mycol Res **94**: 705
61. Loukaci A, Kayser O, Bindseil K-U, Siems K, Frevert J, Abreu PM (2000) New Trichothecenes Isolated from *Holarrhena floribunda*. J Nat Prod **63**: 52
62. Iida A, Konishi K, Kubo H, Tomioka K, Tokuda H, Nishino H (1996) Trichothecinols A, B and C, Potent Anti-Tumor Promoting Sesquiterpenoids from the Fungus *Trichothecium roseum*. Tetrahedron Lett **37**: 9219
63. Okuchi M, Itoh M, Kaneko Y, Doi S (1968) A New Antifungal Substance Produced by *Myrothecium*. Agric Biol Chem **32**: 394; idem (1973) Sesquiterpene Antibiotic, Compound A-2, Minor Component from *Myrothecium*, I. Fermentation, Isolation, Physicochemical Properties, and the Structure. J Antibiot **26**: 562
64. Habermehl G (1989) Isolation and Structure of New Toxins from Plants. Pure Appl Chem **61**: 377

65. Greenhalgh R, Blackwell BA, Savard M, Miller JD, Taylor A (1988) Secondary Metabolites Produced by *Fusarium sporotrichioides* DAOM 165006 in Liquid Culture. J Agric Food Chem **36**: 216
66. Ueno Y, Sato N, Ishii K, Sakai K, Tsunoda H, Enomoto M (1973) Biological and Chemical Detection of Trichothecene Mycotoxins of *Fusarium* Species. Appl Microbiol **25**: 699
67. Sigg HP, Mauli R, Flury F, Hauser D (1965) Die Konstitution von Diacetoxyscirpenol. Helv Chim Acta **48**: 962
68. Gilgan MW, Smalley EB, Strong FM (1966) Isolation and Partial Characterization of a Toxin from *Fusarium tricinctum* on Moldy Corn. Arch Biochem Biophys **114**: 1
69. Hussein HM, Baxter M, Andrew IG, Franich RA (1991) Mycotoxin Production by *Fusarium* Species Isolated from New Zealand Maize Fields. Mycopathologia **113**: 35
70. Cole M, Rolinson GN (1972) Microbial Metabolites with Insecticidal Properties. Appl Microbiol **24**: 660
71. Mirocha CJ, Abbas HK, Kommedahl T, Jarvis BB (1989) Mycotoxin Production by *Fusarium oxysporum* and *Fusarium sporotrichioides* Isolated from *Baccharis* spp. from Brazil. Appl Environ Microbiol **55**: 254
72. Ishii K (1975) Two New Trichothecenes Produced by *Fusarium* sp. Phytochemistry **14**: 2469
73. Ripperger H, Seifert K, Romer A, Rullkotter J (1975) Isolierung von Diacetoxyscirpenol aus *Fusarium solani* var. *coeruleum*. Phytochemistry **14**: 2298
74. Greenhalgh R, Miller JD, Visconti A (1991) Toxigenic Potential of *Fusarium compactum* R8287 and R8293. J Agric Food Chem **39**: 809
75. Ghosal S, Biswas K, Srivastava RS, Chakrabarti DK, Chaudhary KCB (1978) Toxic Substances Produced by *Fusarium*, V: Occurrence of Zearalenone, Diacetoxyscirpenol, and T-2 Toxin in Moldy Corn Infected with *Fusarium moniliforme* Sheld. J Pharm Sci **67**: 1768
76. Scott PM, Harwig J, Blanchfield BJ (1980) Screening *Fusarium* Strains Isolated from Overwintered Canadian Grains for Trichothecenes. Mycopathologia **72**: 175
77. Vesonder RF, Golinsky P, Plattner R, Zeitkiewicz DL (1991) Mycotoxin Formation by Different Geographic Isolates of *Fusarium crookwellense*. Mycopathologia **113**: 11
78. Suzuki T, Kurisa M, Hoshino Y, Ichinoe M, Nose N, Tokumaru Y, Watanabe A (1980) Production of Trichothecene Mycotoxins of *Fusarium* sp. in Wheat and Barley Harvested in Saitama Prefecture. J Food Hyg Soc Jpn **21**: 43
79. Altomare C, Logrieco A, Bottalico A, Mule G, Moretti A, Evidente A (1995) Production of Type A Trichothecenes and Enniatin B by *Fusarium sambucinum* Fuckel sensu lato. Mycopathologia **129**: 177
80. Ishii K, Sakai K, Ueno Y, Tsunoda H, Enomoto M (1971) Solaniol, a Toxic Metabolite of *Fusarium solani*. Appl Microbiol **22**: 718
81. Ueno Y, Ishii K, Sakai K, Kanaeda S, Tsunoda H, Tanaka T, Enomoto M (1972) Toxicological Approaches to the Metabolites of *Fusaria*, IV. Microbial Survey on "Bean-Hulls Poisoning of Horses" with the Isolation of Toxic Trichothecenes, Neosolaniol and T-2 Toxin of *Fusarium solani* M-1-1. Jpn J Exp Med **42**: 187
82. Ueno Y, Sato N, Ishii K, Sakai K, Enomoto M (1972) Toxological Approaches to the Metabolites of *Fusaria*, V. Neosolaniol, T-2 Toxin and Butenolide, Toxic Metabolites of *Fusarium sporotrichioides* NRRL 3510 and *Fusarium poae* 3287. Jpn J Exp Med **42**: 461
83. Harwig J, Scott PM, Stoltz DR, Blanchfield BJ (1979) Toxins of Molds from Decaying Tomato Fruit. Appl Environ Microbiol **38**: 267

84. Ueno Y, Ishii K, Sawano M, Ohtsubo K, Matsuda Y, Tanaka T, Kurata H, Ichinoe M (1977) Toxological Approaches to the Metabolites of *Fusaria*, XI. Trichothecenes and Zearalenone from *Fusarium* Species Isolated from River Sediments. Jpn J Exp Med **47**: 177
85. Desjardins AE, Plattner RD (1989) Trichothecene Toxin Production by Strains of *Gibberella pulicaris* (*Fusarium sambucinum*) in Liquid Culture and in Potato Tubers. J Agric Food Chem **37**: 388
86. Bosch U, Mirocha CJ (1992) Toxin Production by *Fusarium* Species from Sugar Beets and Natural Occurrence of Zearalenone in Beets and Beet Fibers. Appl Environ Microbiol **58**: 3233
87. Abramson D, Clear RM, Smith DM (1993) Trichothecene Production by *Fusarium* spp. Isolated from Manitoba Grain. Canad J Plant Pathol **15**: 147
88. Altomare C, Ritieni A, Perrone G, Fogliano V, Mannina L, Logrieco A (1995) Production of Neosolaniol by *Fusarium tumidum*. Mycopathologia **130**: 179
89. Ilus T, Ward PJ, Nummi M, Adlercreutz H, Gripenberg J (1977) A New Mycotoxin from *Fusarium*. Phytochemistry **16**: 1839
90. Greenhalgh R, Fielder DA, Blackwell BA, Miller JD, Charland J-P, ApSimon JW (1990) Some Minor Secondary Metabolites of *Fusarium sporotrichioides* DAOM 165006. J Agric Food Chem **38**: 1978
91. Baldwin NCP, Bycroft BW, Dewick PM, Gilbert J, Holden I (1985) Biosynthesis of Trichothecene Mycotoxins in *Fusarium culmorum* Cultures. Z Naturforsch **C40**: 514
92. Lee Y-W, Mirocha CJ, Schroeder DJ, Walser MM (1985) TPD-1, a Toxic Component Causing Tibial Dyschondroplasia in Broiler Chickens, and Trichothecenes from *Fusarium roseum* "graminearum". Appl Environ Microbiol **50**: 102
93. Tutel'yan VA, Eller KI, Sobolev VS, Avren'eva LI, Rozynov BV, Bogdanova IA (1984) T-2 Triol – A New Mycotoxin Produced by *Fusarium sporotrichiella*. Dokl Akad Nauk SSSR **274**: 727; Chem Abstr **101**: 3609
94. Lansden JA, Cole RJ, Dorner JW, Cox RH, Cutler HG, Clark JD (1978) A New Trichothecene Mycotoxin from *Fusarium tricinctum*. J Agric Food Chem **26**: 246
95. Bilai VI, Tutel'yan VA, Ellanskaya IA, Eller KI, Sobolev VS (1983) Component Composition of Trichothecenes Produced by *Fusarium sporotrichiella* Bilai. Mikrobiol Zh (Kiev) **45**: 45; (1984) Chem Abstr **100**: 46671
96. Lamprecht SC, Marasas WFO, Sydenham EW, Theil PG, Knox-Davies PS, van Wyk PS (1989) Toxicity to Plants and Animals of an Undescribed, Neosolaniol Acetate-Producing, *Fusarium* sp. from Soil. Plant Soil **114**: 75
97. Jarvis BB, Vrudhula VM, Midiwo JO, Mazzola EP (1983) New Trichoverroids from *Myrothecium verrucaria*: Verrol and 12,13-Deoxytrichodermadiene. J Org Chem **48**: 2576
98. Jarvis BB, Sorenson WG, Hintikka E-L, Nikulin M, Zhou Y, Jiang J, Wang S, Hinkley S, Etzel RA, Dearborn D (1998) Study of Toxin Production by Isolates of *Stachybotrys chartarum* and *Memnoniella echinata* Isolated During a Study of Pulmonary Hemisiderosis in Infants. Appl Environ Microbiol **64**: 3620
99. Bamburg JR, Strong FM (1969) Mycotoxins of the Trichothecene Family Produced by *Fusarium tricinctum* and *Trichoderma lignorum*. Phytochemistry **8**: 2405
100. Kotsonis FN, Ellison RA (1975) Assay and Relationship of HT-2 Toxin and T-2 Toxin Formation in Liquid Culture. Appl Microbiol **30**: 33
101. El-Maghraby OMO, El-Kady IA, Soliman S (1995) Mycoflora and *Fusarium* Toxins of Three Types of Corn Grains in Egypt with Special Reference to Production of Trichothecene Toxins. Microbiol Res **150**: 225

102. Visconti A, Mirocha CJ, Bottalico A, Chelkowski J (1985) Trichothecene Mycotoxins Produced by *Fusarium sporotrichioides* Strain P-11. Mycotoxin Res **1**: 3
103. Corley DG, Miller-Wideman M, Durley RC (1994) Isolation and Structure of Harzianum A: A New Trichothecene from *Trichoderma harzianum*. J Nat Prod **57**: 422
104. Jarvis BB, Midiwo JO, Stahly GP, Pavanasasivam G, Mazzola EP (1980) Trichodermadiene: A New Novel Trichothecene. Tetrahedron Lett **21**: 787
105. Jarvis BB, Pavanasasivam G, Holmlund CE, DeSilva T, Stahly GP (1981) Biosynthetic Intermediates to the Macrocyclic Trichothecenes. J Amer Chem Soc **103**: 472
106. Bean GA, Fernando T, Jarvis BB, Bruton B (1984) The Isolation and Identification of Trichothecene Mycotoxins from a Plant Pathogenic Strain of *Myrothecium roridum*. J Nat Prod **47**: 727
107. Jarvis BB, Salemme J, Morais A (1995) *Stachybotrys* Toxins, 1. Nat Toxins **3**: 10
108. Laurent D, Guella G, Roquebert M-F, Farinole F, Mancini I, Pietra F (2000) Cytotoxins, Mycotoxins and Drugs from a New Deuteromycete, *Acremonium neo-caledoniae*, from the Southwestern Lagoon of New Caledonia. Planta Med **66**: 63
109. Jarvis BB, Vrudhula VM (1983) New Trichoverroids from *Myrothecium verrucaria*: 16-Hydroxytrichodermadienediols. J Antibiot **36**: 459
110. Jarvis BB, Lee Y-W, Comezoglu SN, Yatawara CS (1986) Trichothecenes Produced by *Stachybotrys atra* from Eastern Europe. Appl Environ Microbiol **51**: 915
111. El-Maghraby OMO, Bean GA, Jarvis BB, Aboul-Nasr MB (1991) Macrocyclic Trichothecenes Produced by *Stachybotrys* Isolated from Egypt and Eastern Europe. Mycopathologia **113**: 109
112. Jarvis BB, DeSilva T, McAlpine JB, Swanson SJ, Whittern DN (1992) New Trichoverroids from *Myrothecium verrucaria* Isolated by High Speed Countercurrent Chromatography. J Nat Prod **55**: 1441
113. Jarvis BB, Wang S, Ammon HL (1996) Trichoverroid Stereoisomers. J Nat Prod **59**: 254
114. Gorst-Allman CP, Steyn PS, Vleggaar R, Rabie CJ (1985) Structure Elucidation of a Novel Trichothecene Glycoside Using ^{1}H and ^{13}C Nuclear Magnetic Resonance Spectroscopy. J Chem Soc, Perkin Trans 1: 1553
115. Bekele E, Rottinghaus AA, Rottinghaus GE, Casper HH, Fort DM, Barnes CL, Tempesta MS (1991) Two New Trichothecenes from *Fusarium sporotrichioides*. J Nat Prod **54**: 1303
116. Bamburg JR, Riggs NV, Strong FM (1968) The Structures of Toxins from Two Strains of *Fusarium tricinctum*. Tetrahedron **24**: 3329
117. Vesonder RF, Ellis JJ, Rohwedder WK (1981) Swine Refusal Factors Elaborated by *Fusarium* Strains and Identified as Trichothecenes. Appl Environ Microbiol **41**: 323
118. Burmeister HR, Ellis JJ, Yates SG (1971) Correlation of Biological to Chromatographic Data for Two Mycotoxins Elaborated by *Fusarium*. Appl Microbiol **21**: 673
119. Burmeister HR, Ellis JJ, Vesonder RF (1981) Survey for Fusaria That Produce an Antibiotic That Causes Conidia of *Penicillium digitatum* to Swell. Mycopathologia **74**: 29
120. Yagen B, Joffe AZ (1976) Screening of Toxic Isolates of *Fusarium poae* and *Fusarium sporotrichioides* Involved in Causing Alimentary Toxic Aleukia. Appl Environ Microbiol **32**: 423
121. Ghosal G, Chakrabarti DK, Chaudhary KCB (1976) Toxic Substances Produced by *Fusarium* I: Trichothecene Derivatives from Two Strains of *Fusarium oxysporum* f. sp. *carthami*. J Pharm Sci **65**: 160

122. Soares da Silva N, Kemmelmeir C (1994) Identification of Mycotoxins Produced by *Fusarium graminearum* Isolates Grown on Maize (*Zea mays* L). Arg Biol Tecnol **37**: 551; (1995) Chem Abstr **122**: 284356
123. Eller KI, Sobolev VS, Rozynov BV, Tutel'yan BV (1985) 3'-Hydroxy-T-2 Toxin – A New Mycotoxin from *Fusarium sporotrichiella*. Dokl Akad Nauk SSSR **283**: 1481; Chem Abstr **103**: 210548
124. Kotsonis FN, Ellison RA, Smalley EB (1975) Isolation of Acetyl-T-2 Toxin from *Fusarium poae*. Appl Microbiol **30**: 493
125. Bloem RJ, Smitka TA, Bunge RH, French JC, Mazzola EP (1983) Roridin L-2, a New Trichothecene. Tetrahedron Lett **24**: 249
126. Jarvis BB, Vrudhula VM, Pavanasasivam G (1983) Trichoverritone and 16-Hydroxyroridin L-2, New Trichothecenes from *Myrothecium roridum*. Tetrahedron Lett **24**: 3539
127. Jarvis BB, Wang S, Cox C, Rao MM, Philip V, Varashin MS, Barros CS (1996) Brazilian *Baccharis* Toxins: Livestock Poisoning and the Isolation of Macrocyclic Trichothecene Glucosides. Nat Toxins **4**: 58
128. Chakrabarti DK, Ghosal S (1986) Occurrence of Free and Conjugated 12,13-Epoxytrichothec-9-enes and Zearalenone in Banana Fruits Infected with *Fusarium moniliforme*. Appl Environ Microbiol **51**: 217
129. Bennett GA, Peterson RE, Plattner RD, Shotwell OL (1981) Isolation and Purification of Deoxynivalenol and a New Trichothecene by High Pressure Liquid Chromatography. J Amer Oil Chem Soc **58**: 1002A
130. Yoshizawa T, Morooka N (1973) Deoxynivalenol and Its Monoacetate: New Mycotoxins from *Fusarium roseum* and Moldy Barley. Agric Biol Chem **37**: 2933
131. Vesonder RF, Ciegler A, Jenson AH (1973) Isolation of the Emetic Principle from *Fusarium*-Infected Corn. Appl Microbiol **26**: 1008
132. Vesonder RF, Ciegler A, Jenson AH (1977) Production of Refusal Factors by *Fusarium* Strains on Grains. Appl Environ Microbiol **34**: 105
133. El-Banna AA, Scott PM, Lau P-Y, Sakuma T, Platt HW, Campbell V (1984) Formation of Trichothecenes by *Fusarium solani* var. *coeruleum* and *Fusarium sambucinum* in Potatoes. Appl Environ Microbiol **47**: 1169
134. Bosch U, Mirocha CJ, Abbas HK, di Menna M (1989) Toxicity and Toxin Production by *Fusarium* Isolates from New Zealand. Mycopathologia **108**: 73
135. Abbas HK, Bosch U (1990) Evaluation of Trichothecene and Nontrichothecene Mycotoxins Produced by *Fusarium* in Soybeans. Mycotoxin Res **6**: 13
136. Ramakrishna Y, Bhat RV, Ravindranath V (1989) Production of Deoxynivalenol by *Fusarium* Isolates from Samples of Wheat Associated with a Human Mycotoxicosis Outbreak and from Sorghum Cultivars. Appl Environ Microbiol **55**: 2619
137. Styriak I, Conkova E, Bohm J (1994) Occurrence of *Fusarium sacchari* var. *subglutinans* and Its Mycotoxin Production Ability in Broiler Feed. Folia Microbiol **39**: 579
138. Vesonder RF, Ciegler A, Jenson AH, Rohwedder WK, Weisleder D (1976) Co-Identity of the Refusal and Emetic Principle from *Fusarium*-Infected Corn. Appl Environ Microbiol **31**: 280
139. Tatsuno T (1968) Toxicologic Research on Substances from *Fusarium nivale*. Cancer Res **28**: 2393
140. Golinski P, Vesonder RF, Latus-Zietkiewicz D, Perkowski J (1988) Formation of Fusarenone X, Nivalenol, Zearalenone, α-*trans*-Zearalenol, β-*trans*-Zearalenol, and Fusarin C by *Fusarium crookwellense*. Appl Environ Microbiol **54**: 2147

141. Sugiura Y, Fukasaku K, Tanaka T, Matsui Y, Ueno Y (1993) *Fusarium poae* and *Fusarium crookwellense*, Fungi Responsible for the Natural Occurrence of Nivalenol in Hokkaido. Appl Environ Microbiol **59**: 3334
142. Raza SK, Mallet AI, Howell SA, Thomas PA (1994) An *In-Vitro* Study of the Sterol Content and Toxin Production of *Fusarium* Isolates from Mycotic Keratitis. J Med Microbiol **41**: 204
143. Sundheim L, Nagayama S, Kawamura O, Tanaka T, Brodal G, Ueno O (1988) Trichothecenes and Zearalenone in Norwegian Barley and Wheat. Norwegian J Agric Sci **2**: 49
144. Ghosal S, Chakrabarti DK, Srivastava AK, Srivastava RS (1982) Toxic 12,13-Epoxytrichothecenes from Anise Fruits Infected with *Trichothecium roseum*. J Agric Food Chem **30**: 106
145. Blight MM, Grove JF (1974) New Metabolic Products of *Fusarium culmorum*. Toxic Trichothec-9-en-8-ones and 2-Acetyl-quinazolin-4(3H)-one. J Chem Soc, Perkin Trans 1: 1691
146. Greenhalgh R, Neish GA, Miller JD (1983) Deoxynivalenol, Acetyldeoxynivalenol, and Zearalenone Formation by Canadian Isolates of *Fusarium graminearum* on Solid Substrates. Appl Environ Microbiol **46**: 625
147. Miller JD, Taylor A, Greenhalgh R (1983) Production of Deoxynivalenol and Related Compounds in Liquid Culture by *Fusarium graminearum*. Canad J Microbiol **29**: 1171
148. Ueno Y, Ueno I, Tatsuno T, Ohokubo K, Tsunoda H (1969) Fusarenone-X, a Toxic Principle of *Fusarium nivale* Culture Filtrate. Experientia **25**: 1062
149. Xu Y-C, Huang X, Cai Y-S (1982) Isolation and Structure of CBD_2 – A New Trichothecene Toxin. Weishengwu Xuebao **22**: 35; Chem Abstr **96**: 196245
150. Freeman GG, Morrison RI (1948) Trichothecin: An Antifungal Metabolic Product of *Trichothecium roseum* Link. Nature (Lond) **162**: 30
151. Combrinck S, Gelderblom WCA, Spies HSC, Burger BV, Theil PG, Marasas WFO (1988) Isolation and Characterization of Trichothecin from Corn Cultures of *Fusarium graminearum* MRC 1125. Appl Environ Microbiol **54**: 1700
152. Corley DG, Rottinghaus GE, Tempesta MS (1987) Toxic Trichothecenes from *Fusarium sporotrichioides* (MC-72083). J Org Chem **52**: 4405
153. Yoshizawa T, Shirota T, Morooka NJ (1978) Toxic Substances in Infected Cereals, VI. Deoxynivalenol and Its Acetate as Feed Refusal Principles in Rice Cultured with *Fusarium roseum* No. 117 (ATCC 28114). J Food Hyg Soc Jpn **19**: 178
154. Tatsuno T, Morita Y, Tsunoda H, Umeda M (1970) Recherches Toxicologiques des Substances Metaboliques du *Fusarium nivale*, VII. La Troisieme Substance Metabolique de *F. nivale*, le Diacetate de Nivalenol. Chem Pharm Bull **18**: 1485
155. Otsuka T, Okamoto Y, Hosoya T (1997) Herbicidal F-11073 Manufacture with *Acremonium*. Japanese Patent 09 118682 1997; Chem Abstr **127**: 49311
156. Chatterjee K, Pawlosky RJ, Mirocha CJ, Zhu T-X (1986) Production of Deepoxy-Diacetoxyscirpenol in a Culture of *Fusarium graminearum*. Appl Environ Microbiol **52**: 311
157. Jarvis BB, Midiwo JO, Guo M (1989) 12,13-Deoxytrichoverrins from *Myrothecium verrucaria*. J Nat Prod **52**: 663
158. Bohner B, Fetz E, Harri E, Sigg HP, Stoll Ch, Tamm Ch (1965) Verrucarine und Roridine, 8. Mitteilung. Über die Isolierung von Verrucarin H, Verrucarin J, Roridin D und Roridin E aus *Myrothecium*-Arten. Helv Chim Acta **48**: 1079
159. Smitka TA, Bunge RH, Bloem RJ, French JC (1984) Two New Trichothecenes, PD 113325 and PD 113326. J Antibiot **37**: 823

160. Harrach B, Mirocha CJ, Pathre SV, Palyusik M (1981) Macrocyclic Trichothecene Toxins Produced by a Strain of *Stachybotrys atra* from Hungary. Appl Environ Microbiol **41**: 1428
161. Vittimberga JS, Vittimberga BM (1965) The Muconomycins, II. Muconomycin B, a New Biologically Active Compound. J Org Chem **30**: 746
162. Jarvis BB, Midiwo JO, Bean GA, Aboul-Nasr MB, Barros CS (1988) The Mystery of Trichothecene Antibiotics in *Baccharis* Species. J Nat Prod **51**: 736
163. Eppley RM, Mazzola EP, Highet RJ, Bailey WJ (1977) Structure of Satratoxin H, a Metabolite of *Stachybotrys atra*. Application of Proton and Carbon-13 Nuclear Magnetic Resonance. J Org Chem **42**: 240
164. Rizzo I, Varsavky E, Haidukowski M, Frade H (1997) Macrocyclic Trichothecenes in *Baccharis coridifolia* Plants and Endophytes and *Baccharis artemisioides* Plants. Toxicon **35**: 753
165. Namikoshi M, Kobayashi H, Yoshimoto T, Meguro S, Akano K (2000) Isolation and Characterization of Bioactive Metabolites from Marine-Derived Filamentous Fungi Collected from Tropical and Sub-Tropical Coral Reefs. Chem Pharm Bull **48**: 1452
166. Zurcher W, Tamm Ch (1966) Verrucarine und Roridine, 13. Mitteilung. Isolierung von 2'-Dehydroverrucarin A als Metabolit von *Myrothecium roridum* Tode *ex Fr.*, Gattungstyp bei Fries. Helv Chim Acta **49**: 2594
167. Smitka TA, Bunge RW, French JC, Bloem RJ (1984) 12'-Hydroxyverrucarin J and Isosatratoxin H. U.S. Patent 4436750, 13 Mar. 1984; Chem Abstr **101**: 5555
168. Jarvis BB, Lee Y-W, Comezoglu FT, Comezoglu SN, Bean GA (1985) Myrotoxins: A New Class of Macrocyclic Trichothecenes. Tetrahedron Lett **26**: 4859
169. Jarvis BB, Comezoglu FT, Lee Y-W, Flippen-Anderson JL, Gilardi RD, George CF (1986) Novel Trichothecenes from *Myrothecium roridum*. Bull Soc Chim Belg **95**: 681
170. Croft WA, Jarvis BB, Yatawara CS (1986) Airborne Outbreak of Trichothecene Toxicosis. Atmos Environ **20**: 549
171. Zhang L, Gong B, Hao W (2000) Immunosuppressants Isolated from a *Phoma* sp. 299. Studies on Compounds SIPI 299-B and 299-O. Zhongguo Kangshengsu Zazhi **25**: 9; Chem Abstr **133**: 309033
172. Jarvis BB, Midiwo JO, DeSilva T, Mazzola EP (1981) Verrucarin L, a New Macrocyclic Trichothecene. J Antibiot **34**: 120
173. Tamm Ch, Sigg HP, Harri E (1965) Verrucarine H, J, and K. Belgian Patent 638972, April 21, 1964; Chem Abstr **62**: 8947; idem (1968) Verrucarin K Antibiotic. Swiss Patent 436570, Nov. 15, 1967; Chem Abstr **68**: 67750
174. Tamm Ch: Personal communication
175. Breitenstein W, Tamm Ch (1977) Verrucarins and Roridins, Part 34. Verrucarin K, the First Natural Trichothecene Derivative Lacking the 12,13-Epoxy Group. Helv Chim Acta **60**: 1522
176. White EP, Mortimer PH, di Menna ME (1977) Chemistry of the *Myrothecium* Toxins. In: Wyllie TD, Morehouse LG (eds.) Mycotoxic Fungi, Mycotoxins, Mycotoxicoses, Vol. 1, p. 465. Dekker, New York
177. Kishaba AN, Shankland DL, Curtis RW, Wilson MC (1962) Substances Inhibitory to Insect Feeding with Insecticidal Properties from Fungi. J Econ Entomol **55**: 211
178. Vittimberga BM (1963) The Muconomycins, I. Studies on the Structure of Muconomycin A, a New Biologically Active Compound. J Org Chem **28**: 1786
179. Jarvis BB, Yatawara CS (1986) Roritoxins, New Macrocyclic Trichothecenes from *Myrothecium roridum*. J Org Chem **51**: 2906

180. Matsumoto M, Minato H, Uotani N, Matsumoto K, Kondo E (1977) New Antibiotics from *Cylindrocarpon* sp. J Antibiot **30**: 681
181. Eppley RM, Mazzola EP, Stack ME, Dreifuss PA (1980) Structures of Satratoxin F and Satratoxin G, Metabolites of *Stachybotrys atra*: Application of Proton and Carbon-13 Nuclear Magnetic Resonance Spectroscopy. J Org Chem **45**: 2522
182. Jarvis BB, Stahly GP, Pavanasasivam G, Mazzola EP (1980) Structure of Roridin J, a New Macrocyclic Trichothecene from *Myrothecium verrucaria*. J Antibiot **33**: 256
183. Bata A, Harrach B, Ujszaszi K, Kis-Tamas A, Lasztity R (1985) Macrocyclic Trichothecene Toxins Produced by *Stachybotrys atra* Strains Isolated in Middle Europe. Appl Environ Microbiol **49**: 678
184. Minato H, Katayama T, Tori K (1975) Vertisporin, a New Antibiotic from *Verticinimonosporium diffractum*. Tetrahedron Lett **16**: 2579
185. Namikoshi M, Akano K, Meguro S, Kasuga I, Mine Y, Takahashi T, Kobayashi H (2001) A New Macrocyclic Trichothecene, 12,13-Deoxyroridin E, Produced by the Marine-Derived Fungus *Myrothecium roridum* Collected in Palau. J Nat Prod **64**: 396
186. Eppley RM, Bailey WJ (1973) 12,13-Epoxy-Δ^9-Trichothecenes as the Probable Mycotoxins Responsible for Stachybotryotoxicosis. Science **181**: 758
187. Busam L, Habermehl GG (1982) Accumulation of Mycotoxins by *Baccharis coridifolia*: A Reason for Livestock Poisoning. Naturwiss **69**: 392
188. Jarvis BB, Midiwo JO, Tuthill D, Bean GA (1981) Interaction Between the Antibiotic Trichothecenes and the Higher Plant *Baccharis megapotamica*. Science **214**: 460
189. Whyte AC, Gloer JB, Scott JA, Malloch J (1996) Cercophorins A–C: Novel Antifungal and Cytotoxic Metabolites from the Coprophilous Fungus *Cercophora areolata*. J Nat Prod **59**: 765
190. Jarvis BB, Wang S (1999) Stereochemistry of the Roridins. Diastereoisomers of Roridin E. J Nat Prod **62**: 1284
191. Freckman WG, Jakubowski ZL, Bunge RH, French JC, Balta LE (1984) Antibiotic Roridin E-2. Canadian Patent 1157405, 22 Nov., 1983; Chem Abstr **100**: 137375
192. Kobayashi A, Nakai Y, Kawasaki T, Kawazu K (1989) Fungal Trichothecenes Which Promote Callus Initiation from the Alfalfa Cotyledon. Agric Biol Chem **53**: 585
193. Jarvis BB, Wells KM, Lee Y-W, Bean GA, Kommedahl T, Barros CS, Barros SS (1987) Macrocyclic Trichothecene Mycotoxins in Brazilian Species of *Baccharis*. Phytopathol **77**: 980
194. Jarvis BB, Pena NB, Rao MM, Comezoglu SN, Comezoglu TF, Mandava NB (1985) Allelopathic Agents from *Parthenium hysterophorus* and *Baccharis megapotamica*. In: Thompson AC (ed.) The Chemistry of Allelopathy. Biochemical Interactions. ACS Symposium Series 268, p. 149
195. Samples D, Hill DW, Bridges CH, Camp BJ (1984) Isolation of a Mycotoxin (Roridin A) from *Phomopsis* sp. Vet Hum Toxicol **26**: 21
196. Jarvis BB, Midiwo JO, Flippen-Anderson JL, Mazzola EP (1982) Stereochemistry of the Roridins. J Nat Prod **45**: 440
197. Isaka M, Punya J, Lertwerawat Y, Tanticharoen M, Thebtaranonth Y (1999) Antimalarial Activity of Macrocyclic Trichothecenes Isolated from the Fungus *Myrothecium verrucaria*. J Nat Prod **62**: 329
198. Sugawara T, Tanaka A, Nagai K, Suzuki K, Okada G (1997) New Members of the Trichothecene Family. J Antibiot **50**: 778
199. Jarvis BB, Comezoglu SN, Rao MM, Pena NB, Boettner FE, Williams TM, Forsyth G, Epling B (1987) Isolation of Macrocyclic Trichothecenes from a Large-Scale Extract of *Baccharis megapotamica*. J Org Chem **52**: 45

200. Habermehl GG, Busam L, Heydel P, Mebs D, Tokarnia CH, Dobereiner J, Spraul M (1985) Macrocyclic Trichothecenes: Cause of Livestock Poisoning by the Brazilian Plant *Baccharis coridifolia*. Toxicon **23**: 731
201. Habermehl GG, Busam L, Stegemann J (1984) Miotoxin A: A Novel Macrocyclic Trichothecene from the Brazilian Plant *Baccharis coridifolia*. Z Naturforsch **C39**: 212
202. Habermehl GG, Busam L (1984) Miotoxin B and C, zwei neue macrocylische Trichothecene aus *Baccharis coridifolia* DC. Liebigs Ann Chem: 1746
203. Kupchan SM, Streelman DR, Jarvis BB, Dailey RG, Sneden AT (1977) Isolation of Potent New Antileukemic Trichothecenes from *Baccharis megapotamica*. J Org Chem **42**: 4221
204. Kupchan SM, Jarvis BB, Dailey RG, Bright W, Bryan RF, Shizuri Y (1976) Baccharin, a Novel Potent Antileukemic Trichothecene Triepoxide from *Baccharis megapotamica*. J Amer Chem Soc **98**: 7092
205. Habermehl GG, Busam L, Spraul M (1985) Macrocyclische Trichothecene aus *Baccharis coridifolia*, II. Miotoxin D und Isomiotoxin D, zwei neue macrocyclische Trichothecene aus *Baccharis coridifolia* DC. Liebigs Ann Chem: 633
206. Jarvis BB, Comezoglu SN, Ammon HL, Breedlove CK, Miller RW, Woode MK, Streelman DR, Sneden AT, Dailey RG, Kupchan SM (1987) New Macrocyclic Trichothecenes from *Baccharis megapotamica*. J Nat Prod **50**: 815
207. Zamir LO, Devor KA, Nikolakakis A, Nadeau Y, Sauriol F (1992) Structures of New Metabolites from *Fusarium Species*: An Apotrichothecene and Oxygenated Trichodienes. Tetrahedron Lett **33**: 5181
208. Greenhalgh R, Blackwell BA, Pare JRJ, Miller JD, Levandier D, Meier R-M, Taylor A, ApSimon JW (1986) Isolation and Characterization by Mass Spectrometry and NMR Spectroscopy of Secondary Metabolites of Some *Fusarium Species*. In: Steyn PS, Vleggaar R (eds.) Mycotoxins and Phycotoxins, p. 137. Elsevier, Amsterdam
209. Roesslein L, Tamm Ch, Zurcher W, Reisen A, Zehnder M (1988) Sambucinic Acid, a New Metabolite of *Fusarium sambucinum*. Helv Chim Acta **71**: 588
210. Mohr P, Tamm Ch, Zurcher W, Zehnder M (1984) Sambucinol and Sambucoin, Two New Metabolites of *Fusarium sambucinum* Possessing Modified Trichothecene Structures. Helv Chim Acta **67**: 406
211. Greenhalgh R, Fielder DA, Morrison LA, Charland J-P, Blackwell BA, Savard ME, ApSimon JW (1989) Secondary Metabolites of *Fusarium* Species: Apotrichothecene Derivatives. J Agric Food Chem **37**: 699
212. Corley DG, Rottinghaus GE, Tempesta MS (1987) Secondary Metabolites from *Fusarium*. Two Modified Trichothecenes from *Fusarium sporotrichioides* MC-72083. J Nat Prod **50**: 897
213. Schmidt R, Zajkowski P, Wink J (1995) Toxicity of *Fusarium sambucinum* Fuckel sensu lato to Brine Shrimp. Mycopathologia **129**: 173
214. Ziegler FE, Nangia A, Tempesta MS (1988) Sporol: A Structure Revision. Tetrahedron Lett **29**: 1665
215. Kononenko GP, Bekker AR, Leonov AL, Soboleva NA (1991) Gramilaurone, a Novel Natural Sesquiterpenoid from *Fusarium graminearum* Schw. Tetrahedron Lett **32**: 1893
216. Nozoe S, Machida Y (1970) Structure of Trichodiene. Tetrahedron Lett 2671
217. VanMiddlesworth F, Desjardins AE, Taylor SL, Plattner RD (1986) Trichodiene Accumulation by Ancymidol Treatment of *Gibberella pulicaris*. Chem Commun 1156
218. Desjardins AE, Plattner RD, Beremand MN (1987) Ancymidol Blocks Trichothecene Biosynthesis and Leads to Accumulation of Trichodiene in *Fusarium sporotrichioides* and *Gibberella pulicaris*. Appl Environ Microbiol **53**: 1860

219. Hesketh AR, Gledhill L, Marsh DC, Bycroft BW, Dewick PM, Gilbert J (1990) Isotrichodiol: A Post-Trichodiene Intermediate in the Biosynthesis of Trichothecene Mycotoxins. Chem Commun 1184
220. Patakova-Juzlova P, Rezanka T, Viden I (1998) Identification of Volatile Metabolites from Rice Fermented by the Fungus *Monascus purpureus*. Folia Microbiol **43**: 407
221. Wilkins K (2000) Volatile Sesquiterpenes from *Stachybotrys chartarum*. Environ Sci Pollut Res **7**: 77
222. Zamir LO, Nikolakakis A, Huang L, St-Pierre P, Sauriol F, Sparace S, Mamer O (1999) Biosynthesis of 3-Acetyldeoxynivalenol and Sambucinol. Identification of the Two Oxygenation Steps After Trichodiene. J Biol Chem **274**: 12269
223. McCormick SP, Taylor SL, Plattner RD, Beremand MN (1989) New Modified Trichothecenes Accumulated in Solid Culture by Mutant Strains of *Fusarium sporotrichioides*. Appl Environ Microbiol **55**: 2195
224. Zook M, Johnson K, Hohn T, Hammerschmidt R (1996) Structural Characterization of 15-Hydroxytrichodiene, a Sesquiterpenoid Produced by Transformed Tobacco Cell Suspension Cultures Expressing a Trichodiene Synthase Gene from *Fusarium sporotrichioides*. Phytochemistry **43**: 1235
225. Zamir LO, Devor KA, Morin N, Sauriol F (1991) Biosynthesis of Trichothecenes: Oxygenation Steps Post-trichodiene. Chem Commun 1033
226. Zamir LO (1989) Biosynthesis of 3-Acetyldeoxynivalenol and Sambucinol. Tetrahedron **45**: 2277
227. Zamir LO, Devor KA, Nadeau Y, Sauriol F (1987) Structure Determination and Biosynthesis of a Novel Metabolite of *Fusarium culmorum*, Apotrichodiol. J Biol Chem **262**: 15354
228. Zamir LO, Nikolakakis A, Sauriol F, Mamer O (1999) Biosynthesis of Trichothecenes and Apotrichothecenes. J Agric Food Chem **47**: 1823
229. Fort DM, Barnes CL, Tempesta MS, Casper HH, Bekele E, Rottinghaus AA, Rottinghaus GE (1993) Two New Modified Trichothecenes from *Fusarium sporotrichioides*. J Nat Prod **56**: 1890
230. Nozoe S, Machida Y (1972) The Structures of Trichodiol and Trichodiene. Tetrahedron **28**: 5105
231. Hesketh AR, Bycroft BW, Dewick PM, Gilbert J (1993) Revision of the Stereochemistry in Trichodiol, Trichotriol and Related Compounds, and Concerning Their Role in the Biosynthesis of Trichothecene Mycotoxins. Phytochemistry **32**: 105
232. McCormick SP, Taylor SL, Plattner RD, Beremand MN (1990) Bioconversion of Possible T-2 Toxin Precursors by a Mutant Strain of *Fusarium sporotrichioides* NRRL 3299. Appl Environ Microbiol **56**: 702
233. Flesch P, Voigt-Scheuermann I (1993) Isolation and Identification of Iso-Trichothecin from Cultures of the Fungus *Trichothecium roseum*. Wein Wiss **48**: 15
234. Burgess LW, Forbes GA, Windels C, Nelson PE, Marasas WFO (1993) Characterization and Distribution of *Fusarium acuminatum* subsp. *armeniacum* subsp. nov. Mycologia **85**: 119
235. Nirenberg H (1995) Morphological Differentiation of *Fusarium sambucinum* Fuckel s. st., *F. torulosum* (Berk. et Curt.) Nirenberg comb. nov. and *F. venenatum* Nirenberg sp. nov. Mycopathologia **129**: 131
236. Thrane U, Hansen U (1995) Chemical and Physiological Characterization of Taxa in the *Fusarium sambucinum* Complex. Mycopathologia **129**: 183
237. Gams W (1989) Taxonomy and Nomenclature of *Microdochium nivale* (*Fusarium nivale*). In: Chelkowski J (ed.) *Fusarium* Mycotoxins, Taxonomy and Pathogenicity, p. 195. Elsevier, Amsterdam

238. Hintikka E-L (1977) *Stachybotrys*. In: Wyllie TD, Morehouse LG (eds.) Mycotoxic Fungi, Mycotoxins, Mycotoxicoses, Vol. 1, p. 91. Dekker, New York
239. Jong S-C, Birmingham JM, Ma G (1993) In: Stedmans ATCC Fungus Names, p. 242. Williams and Wilkins, Baltimore
240. Panozishvili K, Zolnikova NY, Vorovkov AV (1972) Verrucarin A from *Dendrodochium toxicum*. Chem Nat Compd (Eng Transl) **8**: 244
241. Panozishvili K, Borovkov AV (1974) Roridin A from *Dendrodochium toxicum*. Chem Nat Compd (Eng Transl) **10**: 408
242. Tulloch M (1972) The Genus *Myrothecium* Tode ex Fr. Commonwealth Mycological Institute, Mycological Papers **130**: 1
243. Miller JD, Greenhalgh R, Wang Y, Lu M (1991) Trichothecene Chemotypes of Three *Fusarium* Species. Mycologia **83**: 121
244. Ichinoe M, Kurata H, Sugiura Y, Ueno Y (1983) Chemotaxonomy of *Gibberella zeae* with Special Reference to the Production of Trichothecenes and Zearalenone. Appl Environ Microbiol **46**: 1364
245. Logrieco A, Bottalico A, Altomare C (1988) Chemotaxonomic Observations on Zearalenone and Trichothecene Production by *Gibberella zeae* from Cereals in Southern Italy. Mycologia **80**: 892
246. Szecsi A, Bartok T (1995) Trichothecene Chemotypes of *Fusarium graminearum* Isolated from Corn in Hungary. Mycotoxin Res **11**: 85
247. Mule G, Logrieco A, Stea G, Bottalico A (1997) Clustering of Trichothecene-Producing *Fusarium* Strains Determined from 28S Ribosomal DNA Sequences. Appl Environ Microbiol **63**: 1843
248. Tamm Ch, Breitenstein W (1980) The Biosynthesis of Trichothecene Mycotoxins. In: Steyn PS (ed.) The Biosynthesis of Mycotoxins. A Study in Secondary Metabolism, p. 69. Academic Press, New York
249. Hanson JR, Marten T, Siverns M (1974) Studies in Terpenoid Biosynthesis, Part XII. Carbon-13 Nuclear Magnetic Resonance Spectra of the Trichothecanes and the Biosynthesis of Trichothecolone from $[2-^{13}C]$-Mevalonic Acid. J Chem Soc, Perkin Trans 1: 1033
250. Achilladelis B, Hanson JR (1968) Studies in Terpenoid Biosynthesis, 1. The Biosynthesis of Metabolites of *Trichothecium roseum*. Phytochemistry **7**: 589
251. Cane DE, Swanson S, Murthy PPN (1981) Trichodiene Biosynthesis and the Enzymatic Cyclization of Farnesyl Pyrophosphate. J Amer Chem Soc **103**: 2136
252. Cane DE, Ha H-J, Pargellis C, Waldmeier F, Swanson S, Murthy PPN (1985) Trichodiene Biosynthesis and the Stereochemistry of the Enzymatic Cyclization of Farnesyl Pyrophosphate. Bioorg Chem **13**: 246
253. Cane DE, Ha H-J (1988) Trichodiene Biosynthesis and the Role of Nerolidyl Pyrophosphate in the Enzymatic Cyclization of Farnesyl Pyrophosphate. J Amer Chem Soc **110**: 6865
254. Cane DE (1989) Stereochemical Studies of Natural Products Biosynthesis. Pure Appl Chem **61**: 493
255. Cane DE (1990) Enzymatic Formation of Sesquiterpenes. Chem Rev **90**: 1089
256. Arigoni D, Cane DE, Muller B, Tamm Ch (1973) The Mode of Incorporation of Farnesyl Pyrophosphate into Verrucarol. Helv Chim Acta **56**: 2946
257. Blackwell BA, Miller JD, Greenhalgh R (1985) ^{13}C NMR Study of the Biosynthesis of Toxins by *Fusarium graminearum*. J Biol Chem **260**: 4243
258. Zamir LO, Nadeau Y, Nguyen C-D, Devor K, Sauriol F (1987) Mechanism of 3-Acetyldeoxynivalenol Biosynthesis. Chem Commun 127

259. Evans R, Hanson JR (1976) Studies in Terpenoid Biosynthesis, Part XIV. Formation of the Sesquiterpene Trichodiene. J Chem Soc, Perkin Trans 1: 326
260. Hohn TM, VanMiddlesworth F (1986) Purification and Characterization of the Cyclase Trichodiene Synthetase from *Fusarium sporotrichioides*. Arch Biochem Biophys **251**: 756
261. Hohn TM, Beremand MN (1989) Regulation of Trichodiene Synthase in *Fusarium sporotrichioides* and *Gibberella pulicaris* (*Fusarium sambucinum*). Appl Environ Microbiol **55**: 1500
262. Cane DE, Chiu H-T, Liang P-H, Anderson KS (1997) Pre-Steady-State Kinetic Analysis of the Trichodiene Reaction Pathway. Biochemistry **36**: 8332
263. Cane DE, Yang G, Xue Q, Shim JH (1995) Trichodiene Synthase. Substrate Specificity and Inhibition. Biochemistry **34**: 2471
264. Cane DE, Shim JH, Xue Q, Fitzsimons BC, Hohn TM (1995) Trichodiene Synthase. Identification of Active Site Residues by Site-Directed Mutagenesis. Biochemistry **34**: 2480; (1997) **36**: 9636
265. Cane DE, Xue Q, Fitzsimons BC (1996) Trichodiene Synthase. Probing the Role of the Highly Conserved Aspartate-Rich Region by Site-Directed Mutagenesis. Biochemistry **35**: 12369
266. Hohn TM, Beremand PD (1989) Isolation and Nucleotide Sequence of a Sesquiterpene Cyclase Gene from the Trichothecene-Producing Fungus *Fusarium sporotrichioides*. Gene **79**: 131
267. Hohn TM, Desjardins AE (1992) Isolation and Gene Disruption of the *Tox5* Gene Encoding Trichodiene Synthase in *Gibberella pulicaris*. Mol Plant-Microbe Interact **5**: 249
268. Proctor RH, Hohn TM, McCormick SP (1995) Reduced Virulence of *Gibberella zeae* Caused by Disruption of a Trichothecene Toxin Biosynthetic Gene. Mol Plant-Microbe Interact **8**: 593
269. Fekete C, Logrieco A, Giczey G, Hornok L (1997) Screening of Fungi for the Presence of the Trichodiene Synthase Encoding Sequence by Hybridization to the *Tri5* Gene Cloned from *Fusarium poae*. Mycopathologia **138**: 91
270. Hohn TM, Plattner RD (1989) Expression of the Trichodiene Synthase Gene of *Fusarium sporotrichioides* in *Escherichia coli* Results in Sesquiterpene Production. Arch Biochem Biophys **275**: 92
271. Cane DE, Wu Z, Oliver JS, Hohn TM (1993) Overproduction of Soluble Trichodiene Synthase from *Fusarium sporotrichioides* in *Escherichia coli*. Arch Biochem Biophys **300**: 416
272. Hohn TM, Ohlrogge JB (1991) Expression of a Fungal Sesquiterpene Cyclase Gene in Transgenic Tobacco. Plant Physiol **97**: 460
273. Desjardins AE, Hohn TM, McCormick SP (1992) Effect of Gene Disruption of Trichodiene Synthase on the Virulence of *Gibberella pulicaris*. Mol Plant-Microbe Interact **5**: 214
274. Proctor RH, Hohn TM, McCormick SP (1997) Restoration of Wild-Type Virulence to *Tri5* Disruption Mutants of *Gibberella zeae* via Gene Reversion and Mutant Complementation. Microbiology UK **143**: 2583
275. Niessen ML, Vogel RF (1998) Group Specific PCR-Detection of Potential Trichothecene-Producing *Fusarium* Species in Pure Cultures and Cereal Samples. System Appl Microbiol **21**: 618
276. Doohan FM, Weston G, Rezanoor HN, Parry DW, Nicholson P (1999) Development and Use of a Reverse Transcription-PCR Assay to Study Expression of *Tri5* by *Fusarium* Species *In Vitro* and *In Planta*. Appl Environ Microbiol **65**: 3850

277. Machida Y, Nozoe S (1972) Biosynthesis of Trichothecin and Related Compounds. Tetrahedron Lett 1969
278. Zamir LO, Gauthier MJ, Devor KA, Nadeau Y, Sauriol F (1989) Trichodiene Is a Precursor to Trichothecenes. Chem Commun: 598
279. Zamir LO, Devor KA (1987) Kinetic Pulse-Labeling Study of *Fusarium culmorum*. J Biol Chem **262**: 15348
280. Gledhill L, Hesketh AR, Bycroft BW, Dewick PM, Gilbert J (1991) Biosynthesis of Trichothecene Mycotoxins: Cell-Free Epoxidation of a Trichodiene Derivative. FEMS Microbiol Lett **81**: 241
281. Hohn TM, Desjardins AE, McCormick SP (1995) The *Tri4* Gene of *Fusarium sporotrichioides* Encodes a Cytochrome P450 Monooxygenase Involved in Trichothecene Biosynthesis. Mol Gen Genet **248**: 95
282. Hesketh AR, Gledhill L, Marsh DC, Bycroft BW, Dewick PM, Gilbert J (1991) Biosynthesis of Trichothecene Mycotoxins: Identification of Isotrichodiol as a Post-Trichodiene Intermediate. Phytochemistry **30**: 2237
283. Kimura M, Kaneko I, Komiyama M, Takatsuki A, Koshino H, Yoneyama K, Yamaguchi I (1998) Trichothecene 3-O-Acetyl-transferase Protects both the Producing Organism and Transformed Yeast from Related Mycotoxins. J Biol Chem **273**: 1654
284. Evans R, Hanson JR, Marten T (1976) Studies in Terpenoid Biosynthesis, Part XVI. Formation of the Sesquiterpenoid Trichothecin. J Chem Soc, Perkin Trans 1: 1212
285. Zamir LO, Devor KA, Nikolakakis A, Sauriol F (1990) Biosynthesis of *Fusarium culmorum* Trichothecenes. The Roles of Isotrichodermin and 12,13-Epoxytrichothec-9-ene. J Biol Chem **265**: 6713
286. Zamir LO, Devor KA, Nikolakakis A, Sauriol F (1996) Biosynthesis of the Trichothecene 3-Acetyldeoxynivalenol: Cell-Free Hydroxylations of Isotrichodermin. Canad J Microbiol **42**: 828
287. Hesketh AR, Gledhill L, Bycroft BW, Dewick PM, Gilbert J (1993) Potential Inhibitors of Trichothecene Biosynthesis in *Fusarium culmorum*: Epoxidation of a Trichodiene Derivative. Phytochemistry **32**: 93
288. Zamir LO, Nikolakakis A, Devor KA, Sauriol F (1996) Biosynthesis of the Trichothecene 3-Acetyldeoxynivalenol. Is Isotrichodermin a Biosynthetic Precursor? J Biol Chem **271**: 27353
289. Udell MN, Dewick PM (1989) Metabolic Conversions of Trichothecene Mycotoxins: De-Esterification Reactions Using Cell-Free Extracts of *Fusarium*. Z Naturforsch **44C**: 660
290. Alexander NJ, Hohn TM, McCormick SP (1998) The *TRI11* Gene of *Fusarium sporotrichioides* Encodes a Cytochrome P-450 Monooxygenase Required for C-15 Hydroxylation in Trichothecene Biosynthesis. Appl Environ Microbiol **64**: 221
291. Zamir LO, Farah CA (2000) Is Fusarium culmorum Isotrichodermin-15-hydroxylase Different from Other Fungal Species? Canad J Microbiol **46**: 143
292. McCormick SP, Hohn TM, Desjardins AE (1996) Isolation and Characterization of *Tri3*, a Gene Encoding 15-O-Acetyl-transferase from *Fusarium sporotrichioides*. Appl Environ Microbiol **62**: 353
293. Brown DW, McCormick SP, Alexander NJ, Proctor RH, Desjardins AE (2001) A Genetic and Biochemical Approach to Study Trichothecene Diversity in *Fusarium sporotrichioides* and *Fusarium graminearum*. Fungal Genet Biol **32**: 121
294. Beremand MN, Desjardins AE (1988) Trichothecene Biosynthesis in *Gibberella pulicaris*: Inheritance of C-8 Hydroxylation. J Industrial Microbiol **3**: 167

295. Alexander NJ, Proctor RH, McCormick SP, Plattner RD (1997) Genetic and Molecular Aspects of the Biosynthesis of Trichothecenes by *Fusarium*. Cereal Res Commun **25**: 315
296. Park JJ, Chu FS (1996) Partial Purification and Characterization of an Esterase from *Fusarium sporotrichioides*. Nat Toxins **4**: 108
297. Beremand MN, VanMiddlesworth F, Taylor S, Plattner RD, Weisleder D (1988) Leucine Auxotrophy Specifically Alters the Pattern of Trichothecene Production in a T-2 Toxin-Producing Strain of *Fusarium sporotrichioides*. Appl Environ Microbiol **54**: 2759
298. VanMiddlesworth F, Beremand MN, Isbell TA, Weisleder D (1990) T-2 Toxin Biosynthesis: Origin of the Isovalerate Side Chain. J Org Chem **55**: 1237
299. Baldwin NCP, Bycroft BW, Dewick PM, Gilbert J (1986) Metabolic Conversions of Trichothecene Mycotoxins: Biotransformation of 3-Acetyldeoxynivalenol into Fusarenone X. Z Naturforsch **41C**: 845
300. Hohn TM, McCormick SP, Desjardins AE (1993) Evidence for a Gene Cluster Involving Trichothecene-Pathway Biosynthetic Genes in *Fusarium sporotrichioides*. Curr Genet **24**: 291
301. Hohn TM, Desjardins AE, McCormick SP (1993) Analysis of *Tox5* Gene Expression in *Gibberella pulicaris* Strains with Different Trichothecene Production Phenotypes. Appl Environ Microbiol **59**: 2359
302. Wuchiyama J, Kimura M, Yamaguchi I (2000) A Trichothecene Efflux Pump Encoded by *Tri102* in the Biosynthetic Gene Cluster of *Fusarium graminearum*. J Antibiot **53**: 196
303. Proctor RH, Hohn TM, McCormick SP, Desjardins AE (1995) *Tri6* Encodes an Unusual Zinc Finger Protein Involved in Regulation of Trichothecene Biosynthesis in *Fusarium sporotrichioides*. Appl Environ Microbiol **61**: 1923
304. Hohn TM, Krishna R, Proctor RH (1999) Characterization of a Transcriptional Activator Controlling Trichothecene Toxin Biosynthesis. Fungal Genet Biol **26**: 224
305. Matsumoto G, Wuchiyama J, Shingu Y, Kimura M, Yoneyama K, Yamaguchi I (1999) The Trichothecene Biosynthesis Regulatory Gene from the Type B Producer *Fusarium* Strains: Sequence of *Tri6* and Its Expression in *Escherichia coli*. Biosci Biotechnol Biochem **63**: 2001
306. Alexander NJ, McCormick SP, Hohn TM (1999) *Tri12*, a Trichothecene Efflux Pump from *Fusarium sporotrichioides*: Gene Isolation and Expression in Yeast. Mol Gen Genet **261**: 977
307. Muhitch MJ, McCormick SP, Alexander NJ, Hohn TM (2000) Transgenic Expression of the *Tri101* or *PDR5* Gene Increases Resistance of Tobacco to the Phytotoxic Effects of the Trichothecene 4,15-Diacetoxyscirpenol. Plant Sci **157**: 201
308. Kimura M, Shingu Y, Yoneyama K, Yamaguchi I (1998) Features of *Tri101*, the Trichothecene 3-O-Acetyltransferase Gene, Related to the Self-Defense Mechanism in *Fusarium graminearum*. Biosci Biotechnol Biochem **62**: 1033
309. Kimura M, Matsumoto G, Shingu Y, Yoneyama K, Yamaguchi I (1998) The Mystery of the Trichothecene 3-O-Acetyltransferase Gene. Analysis of the Region Around *Tri101* and Characterization of Its Homologue from *Fusarium sporotrichioides*. FEBS Lett **435**: 163
310. McCormick SP, Alexander NJ, Trapp SE, Hohn TM (1999) Disruption of *Tri101*, the Gene Encoding Trichothecene 3-O-Acetyltransferase, from *Fusarium sporotrichioides*. Appl Environ Microbiol **65**: 5252

311. Trapp SC, Hohn TM, McCormick S, Jarvis BB (1998) Characterization of the Gene Cluster for Biosynthesis of Macrocyclic Trichothecenes in *Myrothecium roridum*. Mol Gen Genet **257**: 421
312. Jarvis BB, Mokhtari-Rejali N, Schenkel EP, Barros CS, Matzenbacher NI (1991) Trichothecene Mycotoxins from Brazilian *Baccharis* Species. Phytochemistry **30**: 789
313. Bertoni MD, Cabral D (1991) *Ceratopycnidium baccharidicola* sp. nov. from *Baccharis coridifolia* in Argentina. Mycol Res **95**: 1014
314. Rosso ML, Maier MS, Bertoni MD (2000) Macrocyclic Trichothecene Production by the Fungus Epibiont of *Baccharis coridifolia*. Molecules **5**: 345

Melanin, Melanogenesis, and Vitiligo

Shyamali Roy*

Institute of Natural Products, 8, J. N. Roy Lane, Kolkata 700006, India

Contents

1. Melanin.. 132
   1.1. Introduction.. 132
2. Chemistry of Melanin... 134
   2.1. Isolation and Analysis... 134
   2.2. Solubilization... 135
   2.3. Protein Content.. 135
   2.4. Carboxylic and Phenolic Function................................. 136
   2.5. Chemical Degradation... 136
      2.5.1. Reductive Methods.. 136
      2.5.2. Oxidative Methods.. 137
      2.5.3. Pyrolytic Methods.. 137
   2.6. Spectroscopic Studies.. 138
      2.6.1. UV and IR Spectroscopy..................................... 138
      2.6.2. NMR Spectroscopy... 138
      2.6.3. X-Ray Defraction Study..................................... 138
      2.6.4. ESR Study.. 139
   2.7. Structure of Melanin... 139
      2.7.1. Melanin as Homopolymer..................................... 139
      2.7.2. Melanin as Poikilopolymer.................................. 140
      2.7.3. Melanin as Bipolymer....................................... 141
      2.7.4. Biophysical Model of Melanin Structure..................... 141
      2.7.5. Structure of Phaeomelanin.................................. 143
   2.8. Synthesis of Melanin... 143
      2.8.1. Electrochemical Synthesis.................................. 143
      2.8.2. Photochemical Synthesis.................................... 145
3. Characteristic Biophysicochemical Properties of Melanin................ 145
   3.1. Interaction of Melanin with Light................................ 146
      3.1.1. Melanin in UV and Visible Light............................ 146
      3.1.2. Melanin in the Photoprotection of Skin..................... 146
      3.1.3. Melanin as Light Screen in Eyes............................ 147
   3.2. Melanin and Its Redox Function................................... 148
   3.3. Binding Complexation and Medicinal Aspects of Melanin............ 149
   3.4. Use of Melanin for Defence....................................... 150

*E-mail: Shyamali_radha@yahoo.co.in, Fax: 0091-33-834-0000

4. Melanogenesis 150
   4.1. Melanogenesis *in vivo* 150
      4.1.1. Melanocytes 151
      4.1.2. The Characteristics of the Enzyme 152
      4.1.3. Regulation of Melanogenesis 153
         4.1.3.1. Physiological Factors 154
         4.1.3.2. Organic Sulfur Compounds 154
         4.1.3.3. Metal Ions and Other Chemicals 154
         4.1.3.4. Vitamins 154
         4.1.3.5. Hormones 155
         4.1.3.6. Neural Influence 157
         4.1.3.7. Malpighian Cells 157
         4.1.3.8. UV Light 157
   4.2. Melanogenesis *in vitro* 157
      4.2.1. Enzymatic Melanin Synthesis 157
         4.2.1.1. Rearrangement of Dopachrome 158
         4.2.1.2. Polymerization of DHI 159
      4.2.2. Non-Enzymatic Melanin Synthesis: Model Reaction . 161
         4.2.2.1. Udenfriend System: A Model for Mixed Function Oxidase 161
         4.2.2.2. Melanin Formation Under Udenfriend Conditions 162

5. Vitiligo 164
   5.1. Introduction 164
   5.2. Melanocytotoxicity: Antimelanocyte-Antibodies Formation .. 164
      5.2.1. The Immune Hypothesis 165
      5.2.2. The Neural Hypothesis 165
      5.2.3. The Self-Destruction Hypothesis 165
      5.2.4. The Composite Hypothesis 165
   5.3. Chemotherapy of Vitiligo 166
      5.3.1. Psoralens 166
      5.3.2. Psoralen Action and UV Light 166
      5.3.3. Psoralen Action on Melanogenesis 167
   5.4. Abnormal Biochemical Parameters in Vitiligo 168
   5.5. Status of Tryptophan in the Melanogenic System 169
   5.6. A Composite Hypothesis on Vitiligo 171

References 171

1. Melanin

1.1. Introduction

Melanins, pigments of diverse origin and chemical function, have been subjects of interest for a long time (Aristotle, in "Historia Animalia", 315 B.C.). These natural cosmetics of skin, hair, and feathers usually occur in the form of insoluble fine granules in certain dendritic cells

References, pp. 171–185

of the epidermis. The term melanin ($\mu\varepsilon\lambda\alpha\varsigma$ = black) is, however, misleading and confusing since not all biogenetically related pigments are black. The melanogenic enzyme, tyrosinase, is known to catalyze the biosynthesis of not only black but also red to brown, or even yellow pigments, *e.g.* pheomelanins (*208, 266*). On the other hand, the fascinating colors in the feathers of birds, skins of reptiles and fishes and the blue eyes in animals are optical phenomena due to diffraction, light absorption and scattering, interference, that are produced by melanin granules either in combination or not, with other pigmentary colors (*94, 156*) and by complex formation of the granules with heavy metals (*21, 156*).

Skin pigmentation is a natural phenomenon, and has substantial protective, social, and cosmetic significance, and so, melanin and melanogenesis is of obvious interest and importance for researchers. Earlier, it was believed that the colour of black skin was due to insoluble granular pigments derived from bile (*175*). The first concept of the origin and biogenesis of melanin came at the end of the nineteenth century with recognition of tyrosinase, an enzyme capable of transforming tyrosine into a black product in *Russula nigricans* (*9*). Later, a similar enzyme was also found in other plants and in tissues of various invertebrates and vertebrates (*221*). However, investigation of the black products was not an easy task and rather discouraging, due to their rigid behavior towards solubility, either chemically or enzymatically. But melanin research changed and proceeded dramatically with the identification and demonstration of melanocytes, the pigment synthesizing specialized cells, in the epidermis (*10*). It has now been generally accepted that both natural and synthetic melanins are products of a tyrosinase-tyrosine reaction and in modern terminology melanins are described as complex polymeric indole alkaloids derived from 5,6-dihydroxyindole, the latter compound originating from phenylalanin, tyrosine and 5,6-dihydroxyphenylalanin (*70*).

Melanins, although superficially inert, possess some biologically significant physicochemical properties by acting as radical scavengers, ion-exchangers, metal-binder, *etc.* (*55*). Extensive studies of the chemical reactivities of melanin and putative melanin precursors have led researchers into many areas of scientific research the results of which have been reviewed (*55, 70, 92, 123, 184, 211, 259, 264, 265*).

In the present article, an attempt has been made to give not only a brief account of melanin and melanogenesis but also to discuss the present state of research on vitiligo, an idiopathic disease of pigmentary disorder of the skin.

2. Chemistry of Melanin

Melanin pigmentation is determined mainly by two chemically distinct but biologically related types of pigments. One of these consists of the dark insoluble eumelanins and the other of the alkali soluble phaeomelanins, both of them originating from a common precursor, *i.e.* tyrosine. There exist, however, certain hybrid pigments in the epidermal tissues that possess structural features as well as chemical and physical properties of both eu- and phaeomelanins (*210*). Swan (*259*) in an earlier article of this series has presented a comprehensive account of the chemistry of melanin which covered references up to 1974. Since the present paper highlights vitiligo, a very short discussion of melanin chemistry, emphasizing especially the eumelanins, is given here.

2.1. Isolation and Analysis

The most salient point of melanin chemistry is the isolation of the pigment polymer, occurring *in vivo* or produced *in vitro*, as a single chemical compound of definite composition. The intractable nature of the material, the concomitant protein residue, metal ions, and the hydrated state of the sample are the main difficulties in obtaining the pigment in pure form.

Two methods generally used for the extraction of eumelanins from native sources involve removal of all other components of the tissue by prolonged digestion with conc. HCl at room temperature or boiling with 6 M HCl (*70, 184*). However, this drastic treatment of eumelanins in 6 M HCl may cause some changes in pigment structure and composition (*8, 119*). To avoid these difficulties mild processes have been developed, and in some cases, *e.g.* eye-melanin, sepiomelanin, these can be isolated with minimal damage by mechanical separation of the pigment granules followed by a short treatment with 0.5 N HCl at room temperature and extensive sonication in deionized water (*8*). Purification of melanosomes from melanoma tissues by sucrose density gradient ultracentrifugation after digestion with detergents or proteolytic enzymes, *e.g.* pronase and papain, at neutral pH (*210*) is another mild process. Whether these mild treatments can efficiently remove proteins and other impurities bound to the pigment granules is not clear and thus, in keeping with these studies and other limitations, variation has been noticed in the analytical values of melanins from different sources (*211*). In fact, the analyses reflect the average composition of mixed polymers. However, all melanins from animal sources and *in vitro* tyrosinase melanins contain, as a rule, more than

8% nitrogen, and the use of the molar ratio of carbon and nitrogen (C/N) of the pigment is now preferred for characterization purposes (*119*).

2.2. Solubilization

A melanin is considered completely solubilized if the solution does not scatter light. So far only two approaches for solubilization have been successful, one of which involves treatment of the pigment with Solulene 100 (0.1 M solution of dimethyl-n-dodecyl-n-undecyl ammonium hydroxide in toluene; incubation for 2.5 hr at 75°C). The mechanism of this solubilization process is unknown and degradation of the pigment cannot be ruled out (*186, 281*). The second approach involves treatment of naturally occurring melanosomes and synthetic melanins with a dilute solution of H_2O_2 at pH 9–10. Hydrogen peroxide oxidation in mild alkaline media first solubilizes melanin with no obvious structural change followed by decomposition of the excess H_2O_2 (Pt-black, catalase) in the second stage. Melanin precipitates from this solution under acid conditions but readily redissolved in basic media. Such melanin retains the chromatic characteristics of the intact pigment by its color and ESR signal but accounts for only 60–62% of the weight of the starting materials (*281*). Repeated treatment of this precipitated melanin with H_2O_2 results in no further significant weight loss. It appears that melanin, whether native or synthetic, may comprise two major fractions, only one of which is responsible for the color and ESR signal (*281*).

Melanin solubilization provides an opportunity for determination of molecular weights, and using various methods the molecular weights have been found to range between 1100–6000 irrespective of pigment origin (*70, 281*).

2.3. Protein Content

Native melanin is often conjugated with protein through sulfur linkages and thus forms a melanoprotein complex. From the amount of amino acids (numbering 18) removed by hydrolysis, it has been calculated that protein constitutes about 10% of the weight of melanoprotein (*204*). Conjugation could occur either by addition of cystein residues to the pigment polymer to form a sulfur-linked melanoprotein, or by oxidation of a melanosomal protein, containing a terminal tyrosine residue. The presence of sulfur in the protein-free eumelanins and the formation of cysteic acid and taurine by peracetic acid oxidation of sepiomelanin support the former mechanism (*204*).

There is no direct evidence for the alternative mode of conjugation of the protein through a peptide linkage although oxidation of tyrosine containing peptides by tyrosinase was reported, first by Bu'Lock and Harley-Mason (*30*) and later by Yasunobu *et al.* (*284*) and Rosel *et al.* (*225a*) using model peptides.

Benathan and Wyler (*8*) found that native sepiomelanin loses on treatment with hot $6M$ HCl, most of the "proteic" component during the first 15 min, the remaining part being removed in about an hour. Analysis of the solubilized fraction revealed a mixture of amino acids (11%) along with a small amount of glucosamine. The result indicates that a major portion of protein is loosely bound to the pigment granules.

2.4. Carboxylic and Phenolic Function

The ratio of carboxylic and phenolic groups in native melanin has been found to be 1.1 which is reduced to 0.8 after removal of protein with acid (*8, 288*). Significant differences are also noticed between enzymatic (0.5) and autooxidative (1.7) dopamelanin, the latter possessing the highest proportion of carboxylic groups (*8, 288*) and the longer the oxygenation time the higher the carboxyl content as evidenced by acid decarboxylation. According to Swan and Waggott (*257*), during *in vitro* melanogenesis, H_2O_2 is generated, which attacks the 5,6-indolequinone moieties and gives rise to carboxylate pyrrolic units. The same mechanism also seems to be operative *in vivo* (*204*).

2.5. Chemical Degradation

The chemical degradation of both eu- and pheomelanin has been studied, mainly by three methods. Although the yield of products in most experiments is very low and varies significantly for melanins of different origin, the results appear particularly useful in leading to an understanding of the monomeric units present in the pigment structure.

2.5.1. Reductive Methods

Reduction of sepiomelanin in ethanol at 150°C with hydrogen in the presence of a palladium catalyst yielded 5,6-dihydroxyindole (DHI) (*184*), while mild treatment of sepiomelanin or dopamelanin with sodium borohydride in $0.1\,N$ NaOH results in the formation of some

5,6-dihydroxyindole-2-carboxylic acid (DHICA), the latter arising presumably from incorporation of some pigment precursors in the melanin granule (75). Degradation with hydriodic acid (120, 199) was found to be a specific method for identification of phaeomelanins; aminohydroxyphenylalanine, the degradation product identified by HPLC, is characteristic for melanins derived from 5-S-cysteinyldopa. The reaction follows both reductive and hydrolytic processes.

2.5.2. Oxidative Methods

Alkali fusion of sepiomelanin at 300° yielded DHI, DHICA, 4-methylcatechol and 5,6-dihydroxyindole-4,7-dicarboxylic acid identified by paper chromatography (204). These products indicate the presence of indole units in sepiomelanin some of which are linked through positions 4 and 7. The formation of catechol was ascribed to the presence of dopachrome units in the natural pigment (204). Moreover, when the pigment was boiled with 4% aqueous NaOH solution pyrrole-2,3,5-tricarboxylic acid (PTCA) was obtained in about 1% yield. This substance arises presumably from hydrolysis of terminal carboxylated pyrrole units linked through a carbonyl group to the other structural units (184, 204). Similar results were obtained by degradation of other natural and synthetic pigments (184, 204).

A recent study of sodium hydroxide degradation of both natural and synthetic melanins has revealed the formation of two different components; one (more stable under the reaction conditions used) absorbing in the visible region, and a second absorbing in the UV region. It was postulated that the former is composed of stakes of planar monomer units and that the latter represents the "core" of the polymer, providing protective function against harmful UV radiation (70).

Oxidation of sepiomelanin with potassium permanganate gave four pyrrolic acids, pyrrole-2,3,5- and 2,3,4-tricarboxylic acid, pyrrole-2,3-dicarboxylic acid (PDCA), and 2,3,4,5-pyrroletetracarboxylic acid (184, 204). On the other hand, the yield and number of PDCA increased when decarboxylated sepiomelanin was oxidized with permanganate. The origin of these pyrrolic acids was interpreted as resulting from the oxidative breakdown of various types of DHI units in the pigment backbone (184).

2.5.3. Pyrolytic Methods

Natural black (human hair, bovine eyes) and synthetic tyrosine-, dopa-, dopamine-melanins were investigated by Curie point pyrolysis–

gas chromatography–mass spectrometry (79a). The pigments were characterized in terms of the different ratios of the degradation products aromatic hydrocarbons, phenols, catechols, pyrroles, and indoles.

2.6. Spectroscopic Studies

Melanins have been investigated spectrophotometrically by different authors (70, 259) but the results are of little value in elucidating the structure of the pigment molecule.

2.6.1. UV and IR Spectroscopy

The UV-visible range spectrum of melanin shows high absorption but no definite bands. However, using very dilute KBr pellets, the finger print region allowed characterization of some IR bands as being associated with protonation and deprotonation of titrable groups at different pH and binding of iron to various chelating functional groups and permitted comparison of natural and synthetic melanins (70).

2.6.2. NMR Spectroscopy

Chedekel et al. (59a) examined the ^{13}C NMR spectrum of enzymatically produced melanin and identified the benzylic carbon of L-dopa as the C-3 carbon in the dihydroxyindole repeating unit. In melanin formed by autooxidation, however, the C-3 carbons were included in a pyrrolidone as well as a pyrrole ring and a carbonyl carbon. Eumelanins produced in a similar way from DHI showed no presence of carbonyl-containing structural units. These results strongly suggest that the polymerization step involves the 4 and 7 positions of the indole ring. Moreover, NMR studies of melanins can differentiate samples depending on the source and different functional groups present in the eumelanins can be identified (59a).

2.6.3. X-Ray Defraction Study

More detailed information on the structure and dynamic of melanin-polymers were obtained from X-ray defraction studies through the introduction of a high energy resolution technique, i.e. Rayleigh scattering of Mössbauer radiation (RSMR) (70). Although the spectrum is affected by the contribution of water coordinated to the melanin, some informative peaks have been identified. Thus, the main peaks correspond to the average bond lengths (C–C, C=O, C–N) in the monomer units (1.45 Å), to distances between next-nearest neighbors (2.4 Å), to the perpendicular

References, pp. 171–185

interlayer spacing between indole planes (3.4 Å), and to the distances between atoms in adjacent layers occupying different positions in each monomer unit (4.4 Å). The dynamics of the system are typical of a layer structure characterized by large anisotropies in the bonding forces (*70*).

2.6.4. ESR Study

Melanins possess properties of radicals thus exhibiting paramagnetism, and eumelanins were the first biological molecules studied by electron spin resonance (ESR) spectroscopy (*16, 64, 236*). The ESR spectra of both natural and synthetic melanins are very similar and include a featureless signal with a line width of about 4–6 G, g-value 2.004, and no hyperfine coupling. The spin concentration is very low, within the range of $4-10 \times 10^{17}$ spins/g. Semiconductor models and charge transfer complexes through the stacked monomer units of the eumelanin polymer have been proposed to account for such an unusual type of stable radical property (*98a, b, 169*). ESR measurements of hydrated melanin suspensions on melanin polymers at neutral pH reveal that their spin concentrations are reversible, temperature dependent, and largely controlled by an intragranular equilibrium involving quinone, hydroquinone, and semiquinone units in the polymers (*62*). The ESR signals exhibited by these macromolecules have been used successfully to study the photobiophysical properties of melanins (*234*). Further, attempts have been made to correlate the free radical properties of melanins with their chemical structure, biosynthesis, and physiological role in cells and tissues (*236*).

2.7. Structure of Melanin

Another striking feature of melanin chemistry is the structure of the pigment polymer. The two principal approaches employed for elucidating the structure of these complex molecules were (i) biosynthetic and (ii) analytical. Interestingly, while the first approach provided information on the ultimate monomeric precursors of eu- and pheomelanin, the analytical one was significant in developing methods for shedding light on the elemental composition, functional groups, and structural features of both natural and synthetic melanins.

2.7.1. Melanin as Homopolymer

The fundamental work of Raper (*216*) on the *in vitro* enzymatic synthesis of melanin, using tyrosinase, tyrosine and oxygen, was the first synthetic approach investigating melanogenesis and the structure of

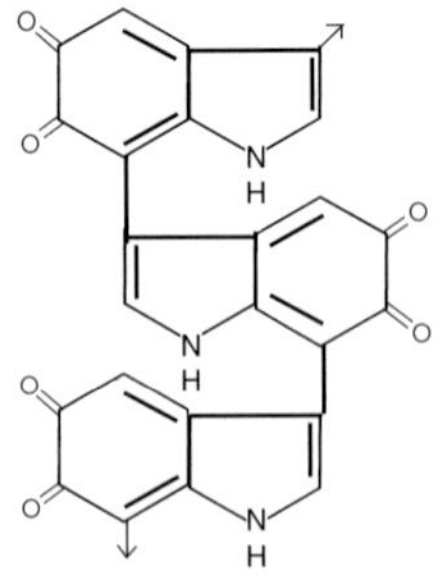

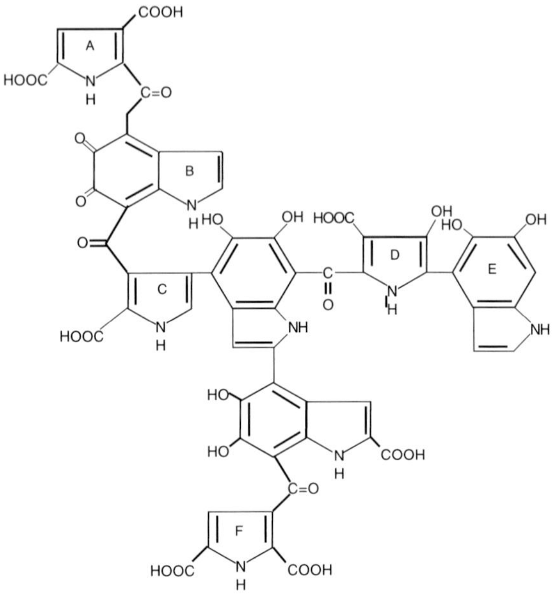

Fig. 1. Partial polymeric structure of eumelanin

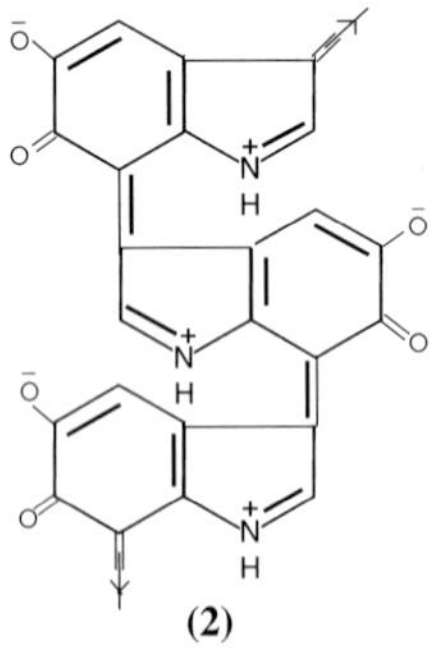

melanin. He concluded that, in the formation of this pigment polymer from 5,6-dihydroxyindole (DHI), quinone formation is a probable first step (*217*). Mason (*163*) found that the oxygen consumption and CO_2 evolution during the oxidative conversion of 5,6-dihydroxyphenylalanin (dopa) to the dopachrome stage were consistent with formation of DHI. He described melanin as a "homopolymer", arising by repeated self-condensation of DHI. Later, in support of this view, it was suggested by other workers (*7, 30*) that for melanin formation from indole-5,6-quinone by a self-condensation process, an unsubstituted 3-position was essential, together with a free position at either C-4 or C-7. A steric-free co-planar structure **1**, with a 3–7 linkage was proposed, the mesomeric form of which with extensive conjugation, **2**, could account for the general light absorption of the polymer. The occasional cross links at the 4–7, 2–4 (or 7) positions, leading to an irregular three-dimensional polymer, might explain the extreme insolubility of melanin (*30*).

2.7.2. Melanin as Poikilopolymer

An analytical approach to studying the structure of melanin, by Nicolaus and co-workers (*184, 204*) led to the concept of melanin as a complex macromolecule or mixture of macromolecules, built from heterogeneous units and linked by more than one type of bonds that are not easily hydrolyzed. The type of linkages, which bind the units, are not known. According to Nicolaus (*184*), an unsubstituted 3-position is not essential, melanin being rather formed by random copolymerization of different highly reactive intermediates formed during the melanization process. The extent to which these monomeric units could contribute to the formation of the pigment polymer, depends upon the chemical and biological environment of the reaction site. Moreover, when free radicals are formed, they could be trapped into the macromolecules and in the end, the irregularity increases further, from the formation of the pyrrol units by the peroxidative cleavage of some labile indole units (Fig. 1) (*259*). The concept of melanin as a "poikilopolymer," was later, confirmed through studies, conducted *in vivo* and *in vitro*, with labelled precursors (*109, 128, 222, 258*).

2.7.3. Melanin as Bipolymer

Swan and Waggott (*257*), on the basis of their various experimental results, suggested that autooxidative dopa-melanin consists in the main of four types of units: 10% of uncyclised units, which are diphenolic (**3**), 10% of indoline-carboxylic acid type (**4**), 65% of indole type (**5**), and 15% of pyrrole type (**6**). Moreover, of the units of types **4** and **5** taken together, one half are quinonoid and the other half are diphenolic; 0.5 H$_2$O was included per average polymer unit.

2.7.4. Biophysical Model of Melanin Structure

The bioelectronic band structure of an indole-5,6-quinone polymer, with an exceptional electron-accepting ability from the lowest empty band in the bonding energy region, was first theoretically calculated by Pullman and Pullman (*212*). The stable free radical properties of the pigment polymers and their role as a sunscreen for biologically harmful quanta (*168*), and their implication in various neurological and psychiatric disorders (*137*), suggested that natural melanins have some significant physiological functions. McGinness (*167*), by applying the theory of Mott (*176*) dealing with the electronic structure of amorphous materials, suggested a model in which the melanin granules act as hypothetical solid state devices that might assume many physiological functions (*168*). Both natural and synthetic melanin act as an amorphous semiconductor threshold switch with a rise in conductivity of melanin under applied voltage. Switching occurs reversibly at potential gradients comparable to gradients existing in some biological systems. The threshold switching in melanins and melanosomes, a rather exotic property of amorphous semiconductors, was studied and according to McGinness *et al.* (*169*) it can occur at biologically attainable electrical field strength. The relation between electronic properties and cellular functions of melanosomes was assumed to be a nonlineal energy transduction device operating by a phonon-electron coupling mechanism (*168*). Phonon-electron coupling plays an important role in amorphous semiconductor theory and according to McGinness *et al.* the coupling of phonons (vibrational modes of macromolecular structure) to an electronically excited state might proceed in both directions and may be particularly efficient in melanin (*169*). It was suggested, further, that the lack of a threshold for ESR signals at low temperature is due to mobility gaps in the amorphous band model as compared with the quantum mechanical model of conduction in amorphous solids (*167, 176*). Evidence for a band model of melanin and its phonon-electron coupling mechanism was also provided by other workers (*69, 173*).

Many biophysical properties of melanin have been studied using these new ideas of melanin structure, *e.g.* the absorption and dispersion of sound waves in melanins. It has been observed that hydrated melanins and melanosomes are exceptional "black" materials with respect to ultrasound absorption (*130*). Melanins are bioelectrets, *i.e.* they can store electrical charge and/or polarization (*25, 81*). Studies on the storage of electrical charge and depolarization in both hydrated natural (epithelium-choroid complexes) and synthetic melanins have revealed a possible role as transconductors for the melanin granules (*25*). Further, melanins have been found in such non-illuminating areas as the brain and the inner ear (*83*) and neuromelanin has been hypothesized to be a component of bioelectronic mechanisms in brain function (*137*). According to McGinness (*170*) phonon-electron coupling may lead to interaction between melanins and pigmented neurons (the fundamental functional units of nervous tissues) which could explain the functional significance of neuromelanin in the brain as well as in the inner ear.

References, pp. 171–185

2.7.5. Structure of Phaeomelanin

Little is known about the structure of phaeomelanins. Although the presence of 1,4-benzothiazine units in these pigment polymers is generally accepted, the other postulates, *i.e.* incorporation of benzothiazole and tetrahydroisoquinoline units into the pigment backbone, need further investigations (59).

2.8. Synthesis of Melanin

Synthetic melanins are obtained by biomimetic oxidation reactions using known precursors. So far, four different methods for melanin synthesis have been reported, *i.e. in vitro* enzymatic, autooxidative, electrochemical, and photochemical methods. Of these, the first two have been generally used, for large scale preparations of the pigment polymers and have been reviewed elsewhere (70, 211). The latter two methods which are discussed here, have been used effectively to understand the mechanism of the melanization process in biological systems.

2.8.1. Electrochemical Synthesis

Electrochemical mechanistic studies of melanin are an outcome of melanin research in 1980. Various authors have employed these methods using various catecholamines and related compounds as substrates (32, 106, 224, 285, 286, 287). These studies have not only confirmed the validity of Raper-Mason's scheme of melanogenesis (see page 158; Fig. 5) but also provided information regarding the mechanism of the chemical steps that occur in the early stages of the melanization process, the identification of each electron-transfer process, and the determination of the rate constants of non-oxidative reactions.

Using a selective amperometric detector in combination with liquid chromatography, the effect of pH on the sequence of events that occur during electrooxidation of catecholamine as well as the quantitative estimation of catecholamines has been studied (129). Thus, the cyclic voltammogram in IM $HClO_4$ (pH range 0.60–6.82 at 15, 20, 25, and 30°C) shows only peaks corresponding to the catechol-quinone redox couple at pH 0.6, since protonation of the amino group prevents the cyclization steps (Fig. 2). At pH 6.36, appearance of a new redox couple indicates the formation of cyclic products, *i.e.* the respective dihydroindole derivative in the reaction mixture. This step is of particular importance in biological systems since the oxidized form of catecholamines is a major factor in determining substrate toxicity that results from

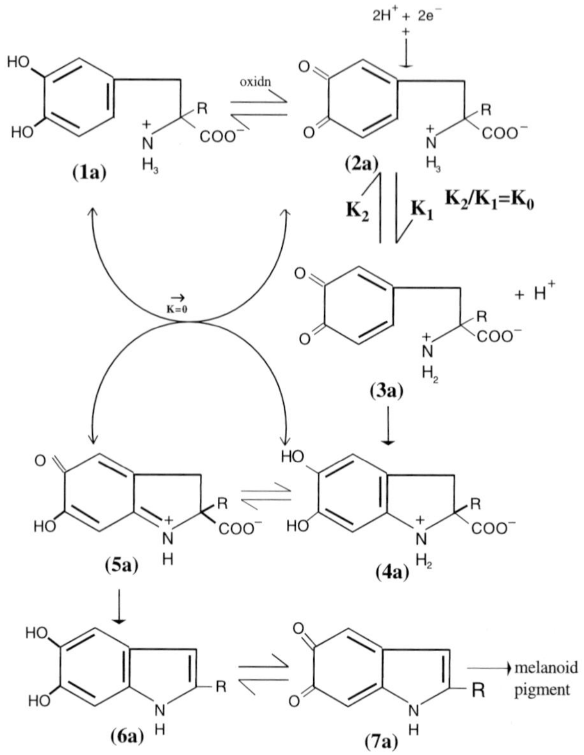

Fig. 2. ECC mechanism of electrochemical oxidation of catecholamines

competitive reactions of the oxidized chatecholamines with the sulfydryl groups of some essential enzymes. Thus the fast cyclizing N-methyl substituted catecholamines are less toxic than the unsubstituted ones which cyclize more slowly (287, 209).

At an even higher pH (>7.68), the absence of a cathodic peak permits an estimate of the half line of the corresponding quinone as being on the order of tens of milliseconds (287); simultaneous darkening around the anode is considered evidence for melanin formation by electrooxidation. The overall reaction sequence in the electrochemical process is, however, very slow involving only a few monolayers.

A theoretical design of an enzymatic chemical mechanism (ECC), both on kinetic evidence and considering its pH dependence, has been suggested for electrochemical oxidation of chatecholamines (32, 224, 286). Thus, α-methyldopa (**1a**) (Fig. 2) first undergoes a two-electron oxidation to α-methyldopa-quinone (**2a** ↔ **3a**), which then cyclizes to

References, pp. 171–185

α-methylcyclodopa (**4a**). A further two-electron exchange (**4a** + **2a** → **5a** + **1a**) then occurs to yield α-methyldopachrome (**5a**) which is fairly stable in solution but is further converted to 5,6-dihydroxy-2-methylindole (**6a**) by a decarboxylative rearrangement.

No direct electron transfer between melanin particles suspended in aqueous buffers and electrodes has been observed. This permitted studies of charge-transfer processes between the chlorpromazine cation radical and catecholamines spectroelectrochemically in order to determine the biological function of chlorpromazine (164).

2.8.2. Photochemical Synthesis

Catecholamines are thermodynamically and photochemically unstable compounds. This property has been utilized in the photochemical synthesis of melanin (74, 277). Thus, a dilute solution of adrenaline, isoprenaline and noradrenaline saturated with oxygen on irradiation (254 nm) gives the corresponding aminochromes in 65, 56, and 35% yield, respectively. Longer irradiation produces melanins (74). Studies of the action spectrum confirmed the excited state of the catecholamine as the primary factor in the transformation processes. N-substituted catecholamines have been found to react more rapidly than the corresponding N-unsubstituted ones (74).

Photooxidation experiments in the presence of azide radicals using pulse radiolysis have revealed that dopa and cysteinyldopa yield first unstable semiquinones which disproportionate to a quinone-quinol complex. The quinones rapidly decay to more stable products, *i.e.* dopaquinones produce dopachromes and cysteinyldopa-quinones rearrange to benzothiazine isomers (136). Further investigations of the oxidation of various melanin precursors, both under physiological conditions (phosphate buffer, pH 7), and in organic solvents, *e.g.* methanol, were performed (77) to study the molecular mechanism of the immediate pigment darkening, *i.e.* natural skin tanning. While experiments in aqueous media showed significant competition between the primary photochemical and auto-oxidative processes, in methanol, a complex mixture of products was obtained.

3. Characteristic Biophysicochemical Properties of Melanin

Natural melanins possess some distinctive physicochemical properties of biological importance. For example, in biological systems, photoprotection of skin against the harmful effect of quanta is believed to be one of the major functions of melanin pigmentation (*197, 198*). In the eye, melanin acts as light screen and strongly resists light adaptation (*81, 93*). Characteristically, the reactions of insoluble melanins are heterogeneous, involving both the surface and the interior of the pigment molecules, thus exhibiting their bifunctional mode of action (*112*). Since the present paper includes a discussion of vitiligo, the etiology of which is still unknown, it is appropriate to discuss briefly some biophysicochemical properties of melanin.

3.1. Interaction of Melanin with Light

3.1.1. Melanin in UV and Visible Light

Melanin shows a relatively structureless spectrum in the ultraviolet and visible range which intensifies with decreasing wavelength (*18*). Basing their arguments on amorphous semiconductor theory, McGinness et al. (*169*) suggested that the absorbed light is not radiated but is instead captured and converted to rotational and vibrational energy (photon-phonon coupling) and that such photon capture is available for any energy level from the UV through the visible and into the IR region. Hence, melanin can be considered black not just in the visible region. An expression for optical density (OD) derived from amorphous semiconductor theory can be stated as follows:

$$ODz^{1,2} = KE_0^{1,2}(Z-1)$$

where Z equals $1.242/\lambda E_0$ (dimensionless), E_0 is the optical bandgap in eV, λ the vacuum wavelength in μm, z a dimensionless independent variable, and K a constant. This relationship appears to hold for both eu- and phaeomelanin.

3.1.2. Melanin in the Photoprotection of Skin

Melanin can effectively protect skin from solar radiation and UV light by dissipating light energy either as heat (*168, 169*) or in a chemical reaction which results in the consumption of molecular oxygen (*232*), or by scavenging active oxygen species, *e.g.* superoxide (*133*), and singlet oxygen (*237*). These phenomena are studied by ESR spectroscopy and spin trapping methods (*133*).

In biological systems, superoxide and H_2O_2 are formed in small quantities by normal processes and both of these species produce harmful effects in tissues. While cell defense mechanisms are adequate to remove these active oxygen species under normal conditions, with exposure to UV light, their concentration increases and thus the function of melanin as an *in situ* quencher is important for the protection of skin (*133*). However, on continuous exposure to UV radiation, melanin itself may become energetically overloaded into a toxic state thus augmenting the radiative damage to cells (*207*). There exists evidence for phaeomelanin's role in sunlight-induced skin cancer (*57, 58*).

The process of photoinduced interaction of melanin with molecular oxygen is accompanied by photobleaching (*232*), and phaeomelanin has been found to be more photolabile and susceptible to photodegradation on prolonged photolysis (*58*).

3.1.3. Melanin as Light Screen in the Eyes

The melanin granules of the eye generate free radicals when irradiated with visible light. The rapid generation of free radicals in light and their subsequent decay in the dark, coupled with the proximity of the melanin granules in the rods and cones of the eye, suggests that melanin plays a role in the visual process which is more important than the mere absorption of stray light (*65*).

The stimulation of a visual receptor in a vertebrate eye by an intense flash of light generates a fast electrical response, the early receptor potential (RP). A similar response in the pigment epithelium-choroid complex (PE–CC) of the eye was observed by Brown (*26*) which unlike the early RP is photostable and resistant to light adaptation. The PE–CC consists of the cell layers immediately behind the retina which are densely pigmented with melanin. Brown concluded that the photopigment involved in the PE–CC response was related to the visual pigments.

Ebrey and Cone (*81*) reexamined the phenomenon using isolated PE–CC layers and found that the amplitute of the major peak of the PE–CC response was proportional to the energy of the stimulus flash. They also studied the action spectrum of the PE–CC response using whole-eye preparations and found that the early RP which is photolabile to continuous light dominated the entire response recorded in the spectrum. At the far end, when steady light fully bleaches the visual pigments, the spectrum of the whole-eye response changes shape and a new response is observed with a flat action spectrum which is similar to that obtained from isolated PE–CC. The authors concluded that the photopigments for these responses absorb all the wavelengths of light equally well, and that the melanins present in the PE–CC cells are primarily responsible for generating at least part of the PE–CC response. This new electrical response of the PE–CC is, however, fundamentally different from the early RP which depends on visual pigments.

3.2. Melanin and Its Redox Function

One of the most characteristic properties of melanins is their ability to exist in both the oxidized quinone and reduced quinol forms by acting either as electron acceptors or electron donors in reaction with reducing and oxidizing agents respectively, thus exhibiting their dual functionality (*33*). These processes merely involve the reversible exchange of two electrons and two protons (*33, 112, 138b*).

To examine changes in the oxidation state of melanins, ESR techniques have been used extensively while to determine the concentration changes of the reagents (oxidants, reductants), spectrophotometric or electrochemical methods have been used (*236*). Owing to the presence of carboxyls and phenolic groups in melanins, positively charged reagents have been found to react much faster than anions or neutral species, especially in basic media (*138a*). Generally reduction of both synthetic and natural melanins results in a lighter color and changes in the ESR spectra. The populations and the role of semiquinone states, assumed to be responsible for the characteristic ESR signals, have been studied for all types of melanins (*236*).

Both reduction and oxidation processes have been found to be biphasic. Thus, in kinetic studies of the reduction of synthetic D,L-dopa melanin with Ti^{3+} and oxidation with Fe^{3+}, respectively, a fast electron-exchange step was followed by a slow second step (*112*). The biphasic character of the electron-exchange processes has been interpreted as being due to a difference in the reaction mechanisms involving the surface and the core of the melanin granules (*112*). Using the oxidation-reduction capacities obtained for the fast electron-exchange processes, one-fourth of the indole units were found at the particle surface. Assuming the same fast rate of electron exchange in both the oxidation and reduction, respectively, the slow diffusion of the reagent (Ti^{3+}, Fe^{3+}, and H^+) in and out of the melanin particle is believed to control the rate of the second phase (*112*).

The electron-exchange properties of melanins have been studied with a number of especial reagents in order to understand the reaction mechanism as well as the role of melanin redox properties in biological systems. The processes have been found to be strongly irradiation dependent (both by visible and UV light). Thus, nitroxide radicals are reversibly reduced by melanins in the dark, and the redox equilibria are altered on irradiation (*233*). Similarly, other processes in living systems (especially the processes involving dihydronoradrenaline, dihydrophenylnoradrenaline, and cytochrome) have been found directly linked to the redox properties of melanins (99).

Of particular significance to biological systems is the reaction of melanins with oxygen and the effect of external factors on this reaction. Thus, pH, illumination with visible light, temperature, and catalase have been studied in detail. Melanins were studied in their native, reduced, oxidized, and methylated forms and the reactions were monitored *via* ESR (231). The rates of oxygen uptake were, generally, higher with illumination and over the pH range 5.5–11.9. In addition to direct electron-exchange properties, melanins act as electron-transfer agents. Thus, some synthetic melanins have been found to accelerate the oxidation of dihydronoradrenaline with $Fe(CN)_6^{3-}$ (99).

3.3. Binding Complexation and Medicinal Aspects of Melanin

Metal cations and organic species carrying positive charges possess strong affinity for melanins (139, 140), and binding complexation which occurs through ion-exchange and/or hydrophobic interaction mechanisms have biological significance (17, 278). These ESR characteristic phenomena have been studied to explain medicinal aspects which may involve the affinity and binding of various molecules to melanins, such as the toxicity of drugs, the malignancy of melanoma cells, and *substantia nigra* (141, 158, 181). For example, a pathological pigmentation often occurs in the skin of the patients taking large doses of chlorpromazine (5). It has been suggested that chlorpromazine acts on the autonomic nervous system by blocking the production of pigment-lightening factors such as melatonin (283). According to some authors, white light produces free radicals from chlorpromazine, and the formation of a stable charge-transfer complex with melanin is the explanation for this chemical hyperpigmentation (5, 141).

Studies of metal binding by both native and synthetic melanins have provided data that not only confirm the ion-exchange properties of melanins but also reveal the relative affinities of metals for melanins (139). Thus, similar to other ion-exchangers, the affinity for melanin increases with the valency of the cation and atomic number of the element (*i.e.* $Cs^+ \gg Li^+; Ba^{2+} \gg Mg^{2+}$). However, the exceedingly high affinity found for Pb^{2+} when compared with similar divalent ions suggests the possible involvement of other factors though protein present in native melanins plays a minor role in the binding of metals (139).

Binding studies combined with ESR spectroscopy have revealed the mechanism of the interaction of metal ions with melanins which indicates the formation of a chelate complex between di- and trivalent diamagnetic metal ions and o-semiquinone radical centers on the pigment polymer (*85, 124*). This interaction often results in an increase of total free radical concentration. Furthermore, the binding capacity varies for melanins of different origins with the number of reactive sites (*230*).

The cation exchange mechanism has also been observed in the binding of organic molecules (mostly bases and often positively charged) to melanins (*140*). A systematic structure-affinity study was reported for a series of heterocyclic compounds and synthetic D,L-melanin. The relative affinities measured from absorption in pH 7 phosphate buffer, showed that π-electron system, basicity, and steric factor are the main determining factors for the affinity towards melanin (*113, 140*). These results are consistent with the stability of charge-transfer complexes between the organic molecule (as donor) and melanin (as acceptor). This structure-affinity relationship can be used for the development of drugs which may selectively target melanocytes (*e.g.* melanoma cells), or drugs with low toxicity that are not accumulated in melanin-containing tissues, such as the eyes.

3.4. Use of Melanin for Defence

The melanin of sepia is found in the ink sac of cuttlefish. Aristotle in his book "Historia Animalia" mentioned that "instead of intestines molluscs have an organ known as the *mytin* where a black substance is found which is especially abundant in cuttlefish; they put forth this black substance when frightened, particularly the cuttlefish". The ink has considerable staining power, and is alkaline and odorless. When fresh it disperses readily in water and is partly soluble in alkali, but insoluble in acid. This biological function of melanin appears to be defensive but the mechanism is uncertain (*270*). However, like other pigments, melanin can be utilized for camouflage purposes in insects and fishes.

4. Melanogenesis

4.1. Melanogenesis *in vivo*

Epidermal melanin synthesis is a multistage process involving fast or slow reactions, some of which are enzyme catalyzed, others requiring only oxygen. The process is controlled by pH, temperature, redox

References, pp. 171–185

potential, activity of the enzyme and the presence of inhibitors. The overall reaction occurs, within a genetically controlled biochemical environment, in specialized cells called melanocytes. Apart from melanin granules, melanocytes contain unique organelles, "premelanosomes" and "melanosomes" in which the biosynthesis of melanin occurs (*89*).

4.1.1. Melanocytes

Melanocytes are dendritic cells which are wedged between the basal cells of the epidermis. They derive from neural crest (*i.e.* at a high point of action of the nervous system) and have complex structures mainly made up of protein and various oxydase systems, including tyrosinase (*160*).

The biosynthesis of melanin has been studied by electron microscopy in epidermal melanocytes after *in vivo* ultraviolet irradiation with either dopa or tyrosine as substrate (*114*). The melanogenic enzyme, originating from minute polypeptide particles (50–100 Å), is transferred into the Golgi area where it condenses to membrane-limited vesicles to produce "protyrosinase". There, protyrosinase is incorporated into a structural protein matrix containing filaments that have a distinctive periodicity. The unit represents a "premelanosome" (*95, 123*), which possesses intense enzyme activity but contains no melanin (*271*). The protyrosinase then becomes active (*i.e.* becomes tyrosinase), biosynthesis of melanin begins, and the unit is called a "melanosome". In their development stages these melanosomes move from the cytoplasm of the melanocyte into dendritic processes (*i.e.* branching protoplasmic processes that produce impulses toward the body of a nerve cell); their melanin content increases while their enzyme activity decreases (*89, 267*), and at the end, when the melanosomes no longer possess enzyme activity, melanin is formed entirely by nonenzymatic polymerization, fills the entire organelle and obscures its (melanin) internal structure. Melanin is transferred to the basal dendritic cells, the keratinocytes. Each melanocyte supplies several keratinocytes with melanins (1:36) thus forming with them an epidermal melanin unit (*89*).

Morphologically, melanocytes and melanin granules from various sources are very similar. However, a definite correlation exists between the size and morphology of the melanocytes of human skin, and their pigment forming activities, *e.g.*, dark color skin contains larger and uniformly more highly reactive dendritic melanocytes than caucasoid skin (*88, 255*), whereas the melanocytes of caucasoid skin, when not exposed to sunlight, are highly variable in dopa reactivity (*213*).

Although the melanosome theory has been accepted in general, some authors support a mitochondrial origin of the melanin granules (*282*).

The chemical compositions of melanosoma and mitochondria have been determined and compared, using differential and density-gradient techniques (*241*). The data show that melanosomal fractions contain a high percentage of Zn and a low content of RNA and phospholipid-P as compared with mitochondrial fractions. The lack of succinoxidase and glutamate oxidase activities support the idea that melanosomes are subcellular particles, different from mitochondria, and contain a particular specialized metabolic pathway through which tyrosine or dopa is converted into melanin (*239*).

4.1.2. The Characteristics of the Enzyme

The traditional view of enzymatic melanogenesis, expressed by different authors (*88*, *145*, *242*), holds that tyrosinase is the melanogenic enzyme, and studies on *in vitro* melanogenesis using mushroom tyrosinase have been considered as valid for mammalian melanogenesis as well, although the enzymes isolated from different sources show qualitative differences (*27*, *126*).

Tyrosinase is a copper-containing glycoprotein that carries a coupled binuclear copper active site capable of catalyzing two distinct reactions: (i) hydroxylation of tyrosine to dihydroxyphenylalanin, *i.e.* dopa (cresolase activity), and (ii) subsequent two-electron oxidation to dopaquinone (catecholase activity) (*221*). Both reactions require oxygen and the enzyme, *i.e.* tyrosinase, in reduced cuprous form. Because of these two catalytic functions, tyrosinase is in modern terminology referred to as a "mixed function oxidase" (*157*). A mechanism (Fig. 3) for hydroxylation

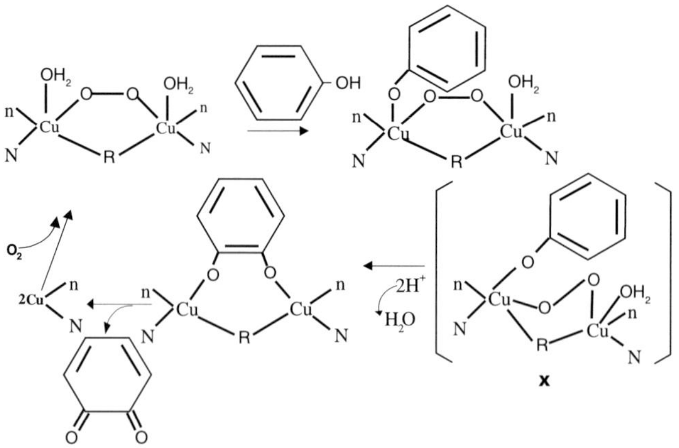

Fig. 3. Proposed mechanism of tyrosinase oxidation

References, pp. 171–185

and oxidation of phenolic substrates to *o*-quinones has been proposed which states that during the *o*-hydroxylation process an intermediate complex X is formed which involves one molecule of oxygen and two neighboring cuprous atoms attached to a protein chain. The formation of such a structure depends on the relative position of the copper atoms, hence the "cresolase activity" would be sensitive to any variation in the configuration of the protein chain (*221, 274*).

The unique property of this enzyme is that the product of the first monoxygenation step, *o*-diphenol, serves as the electron donor for the reduction of the cupric ions with formation of the corresponding *o*-quinone (*161, 274*). The products formed by catecholase activity from tyrosine are extremely reactive and undergo intermolecular reactions to form indole derivatives which subsequently polymerize to melanin.

The tyrosine-tyrosinase reaction is characterized by a variable lag period (*145*), which can be shortened by addition of a catalytic amount of dopa or related compounds. However, there is no time lag with dopa as substrate, and it seems very likely that dopa induces a conformational change in normal tyrosinase and activates the reduction of tyrosinase, thereby acting as a cofactor in the melanization process (*145*). When tyrosinase is present in low concentration as in nonirradiated skin, this lag period is prolonged markedly, whereas in skin exposed *in vivo* to ultraviolet light (*87*), in epidermal sheets (*260*), and in hair bulbs (*88*), the tyrosine-tyrosinase reaction is readily detectable. The presence of tyrosinase in human epidermis was established in 1950 (*86*).

Okun and his associates (*187*) suggest that "peroxidase" rather than copper-dependent tyrosinase, mediates the conversion of tyrosine to melanin in the presence of dopa as cofactor. However, some results obtained in the course of investigations on melanogenesis by other authors contradict their concept (*111, 280*).

Tyrosinase activation has been found to be inhibited by its own substrate *in vitro* and this inhibition-mechanism has been studied (*110, 125, 268*).

4.1.3. Regulation of Melanogenesis

Biochemical analyses of the regulation of pigmentation and proliferation have largely been confined to the population of melanoma cells grown in culture. These studies have revealed that in addition to these tyrosinase-calalyzed steps (*200*) various non-melanosomal regulatory factors are involved in the pathway for melanin biosynthesis (*244*). A short discussion of these factors is given in the following.

4.1.3.1. Physiological Factors

Elevation of temperature increases tyrosinase activity, shortens the induction period and stimulates melanin formation within physiological limits. The enzyme activity is optimal in the pH range 6.7–7.2 during *in vitro* reactions (*183*). The temperature responses of melanin-stimulating hormones have also been studied (*226*).

4.1.3.2. Organic Sulfur Compounds

Organic sulfur compounds are known to inhibit *in vitro* melanin synthesis (*209*), and it was suggested that thiol groups form strong bond with copper, inactivate tyrosinase and thus inhibit melanin formation (*91, 225*). It has now been confirmed that dopaquinone, generated *in situ* by tyrosinase catalyzed oxidation of dopa, reacts with cysteine forming cysteinyldopas (*117*). Hyperpigmentation after UV irradiation proceeds by a marked decrease in glutathione reductase activity and a decrease in reduced glutathione which is an inhibitor of melanin synthesis (*118*). In fact, glutathione reductase activity and reduced glutathione are lower in black than in caucasian skin (*120*). Furthermore, γ-glutamyl transpeptidase was found to be inactivated in the course of melanogenesis (*120*).

4.1.3.3. Metal Ions and Other Chemicals

Metals play an important role in melanin biosynthesis (*244*) and incorporation of various metal-ions into melanins, produced *in vivo* and *in vitro*, respectively, has long been known (*28*). During melanin formation under anaerobic conditions, certain metals (*e.g.* Zn, Fe, Mn *etc.*) catalyzed non-decarboxylating rearrangement of dopachrome (*190*). According to Chakraborty *et al.* (*50*), though autooxydation of dopa is a copper catalyzed reaction, it can equally be catalyzed by nickel, cobalt, and slightly by selenium also. However, lead inhibits melanogenesis (*263*).

Hydroquinone and its derivatives *p*-hydroxypropiophenone, pyridine derivatives, catechol, mercaptamine derivatives, etc., inhibit melanin formation by blocking tyrosinase activity (*56, 61*).

4.1.3.4. Vitamins

Deficiency of pantothenic acid causes depigmentation while administration of folic acid increases hepatic storage of pantothenic acid as the Vitamin B-complex acts synergistically with folic acid in normal pigmentation (*269*). Ascorbic acid maintains optimum sulfhydryl levels in the body and keeps melanin in reduced form (*262*). Any change in the dose levels of this acid (higher or lower), inhibits or accelerates tyrosinase activity (47). Thiamine increases the lag time for induction of the tyrosine-tyrosinase reaction and disrupts the equilibrium of dopa-dopaquinone autooxidation (261).

It has now been confirmed that cholecalciferol, Vitamin D_3, has a stimulatory effect on the melanocytes, due to the *de novo* synthesis of tyrosinase and transfer of melanin granules to the surrounding keratinocytes with elongation of their dendrites (269).

4.1.3.5. Hormones

The pituitary gland plays an important role in the hormonal control of melanin pigmentation (142). The highly reactive melanocyte stimulating hormone intermedine (MSH) secreted by this gland is not only able to darken the human skin (150), but has numerous biological activities in higher vertebrates. For example, it acts on the thyroid gland of the rabbit (68, 235) and on the blood-aqueous humor barrier in the rabbit eye (80); α-MSH acts on the adrenal cortex (102), β-MSH acts on the central nervous system *etc*. (134, 154). The hormone adrenocorticotropin (ACTH) secreted by the anterior lobe of the pituitary gland disperses melanin granules in the frog-melanophore and moderately stimulates epidermal melanocytes (171).

Tyrosinase activity decreases after hypophysectomy, but can be restored by chronic administration of pituitary hormones like prolactin, MSH, and ACTH to a greater or lesser extent (131). Prolactin can promote *in vivo* melanin synthesis not only by stimulating tyrosinase activity, but also by increasing the supply of available substrate (131). However, it is still unknown which of the pituitary hormones is responsible for hyperpigmentation in Addison's disease in man.

Melatonin (*N*-acetyl-5-methoxytryptamine) is a pineal gland hormone which aggregates melanin granules in the dermal melanophores and is capable of lightening the color of frog melanocytes by reversing the darkening action of MSH, ACTH, and caffeine (116, 149). Its effect on the human and other mammalian melanocytes is, however, questionable (253).

Very small amounts of adrenalin and noradrenalin inhibit MSH action on skin pigmentation, but there is no direct evidence of their action on mammalian melanocytes (151, 174).

Among sex hormones, the female hormones are strong stimulants of melanogenesis (251). The reports reveal that administration of small doses of estrogen to ovariectomized guinea pigs increases the melanin content of melanocytes in all skin regions examined, while the effect of large doses on skin pigmentation is direct, and not *via* the pituitary gland (*11, 250*). Small doses of progesterone slightly stimulate melanogenesis (*252*), while very large ones reduce the amount of free melanin (*250*). Interestingly, a mixture of estrogen and progesterone (in small doses), when given to ovariectomized guinea pigs, causes greater stimulation of melanogenesis than one produced by either hormone given alone in the same doses (*11*). The significance of sex hormones in the tanning of the skin of women has been examined by Hamilton *et al.* (*104*). Testosterone has been shown to increase skin pigmentation in the castrated and hypogonadal human male and in some other species (*10, 73*).

Experiments with cyclic-AMP have revealed that in addition to an increase in melanin formation by increasing tyrosinase activity, it can also enhance the pigment transfer (*115, 276*). Dibutyryl-cyclic-AMP has been found to induce melanin formation in the hair bulb of chinchilla mice (*115*). The roles of prostaglandin and cyclic nucleotides, such as cyclic-GMP, remain unclear for men at present.

The hormonal control on pigmentation may be summarized in Fig. 4.

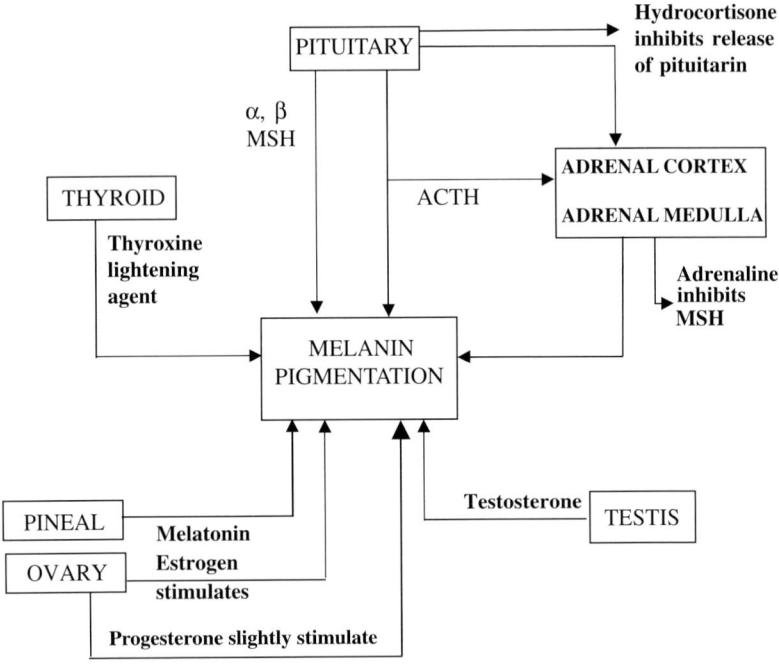

Fig. 4. Hormonal control of melanogenesis

References, pp. 171–185

4.1.3.6. Neural Influence

Although melanocytes are derived developmentally from the neural crest, the neural influence on human skin pigmentation is not clearly understood (*152*). According to some authors, vitiligo results from an increase in activity of the peripheral nerve endings in the skin (*24*). Neural influences of pigmentation have also been noticed in lower vertebrates where skin color changes rapidly (*102, 194*).

4.1.3.7. Malpighian Cells

Melanocytes synthesize melanins and malpighian cells help in the transfer and distribution of the melanin granules in the epidermis, thus operating together as a single functioning unit in the process of skin pigmentation (*79, 89*). According to Fitzpatrick *et al.* (*89*), malpighian cells control the rate of production of melanin granules by melanocytes, *i.e.* the production of melanin granules in melanocytes and their pool of malpighian cells.

4.1.3.8. UV Light

The role of UV light on the pigmentary system has been reported in detail by Quevedo (*214*). It has been documented that if a single exposure to UV light is administered and delayed tanning is studied, then increased enzyme activity, increased pigment formation, increased number of melanocytes and melanosomes, increased degree of melanization, and increased pigment transfer occur (*100, 198*). The regulatory events at the operon nucleic acid level are, however, not yet clear.

4.2. Melanogenesis *in vitro*

The chemistry of melanins reveals that it is difficult to study these pigment polymers using conventional chemical and physical methods. On the other hand, by using known precursors biosynthetic studies under biologically relevant conditions can be used successfully to understand the chemistry of melanin, *i.e.* the identification of the intermediates, their reactivities, the formation of melanin, and its properties.

4.2.1. Enzymatic Melanin Synthesis

This biosynthetic approach was first exploited with remarkable success by Raper (*216*). Later, using advanced research techniques, biosynthetic studies of eumelanin polymers have revealed many facts regarding

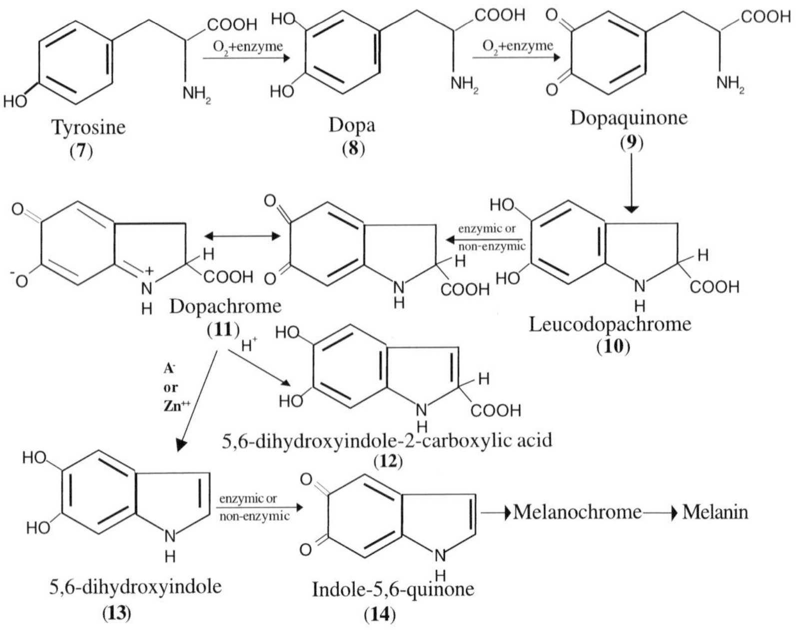

Fig. 5. The Raper-Mason scheme of melanogenesis

the diversity of the pigment origin, the heterogeneity of its structure, and its unique properties.

In the Raper-Mason scheme of melanin biosynthesis (Fig. 5) (*216, 217*), tyrosine is enzymatically converted *via* dopa to dopaquinone. The subsequent oxidation steps leading to melanin formation depend upon the biochemical environment of the reaction site. However, the melanization process *in vitro* or *in vivo* has two important features; the rearrangement of dopachrome and the oxidative polymerization of 5,6-dihydroxyindoles leading to melanochrome.

4.2.1.1. Rearrangement of Dopachrome

For the rearrangement of dopachrome, one possible mechanism which has been accepted by different authors (*66, 218, 255*), involves a hydrogen shift from position 3 and the formation of a quinone methide followed by subsequent decarboxylation to **13**. Analysis of the analytical data from various laboratories (*66, 108, 202*) has shown that the yield of DHI *vs* DHICA is about 95 to 5, at a pH range from 3 to 8.5 which indicates that tyrosinase-catalyzed synthetic dopamelanin is made up mainly of DHI-derived units, as proposed by Mason (*163*). Hence, the carboxyl content of dopamelanin must be due to peroxidative cleavage of indole units rather than to a random incorporation of carboxylated biosynthetic intermediates, a view which contradicts the concept of the poikilopolymer model of melanin structure (259).

The presence of certain metal ions, especially Cu, Fe, Zn *etc.*, in melanin biosynthesis, however, accelerates the non-decarboxylating rearrangement of dopachrome leading to the formation of DHICA rather than DHI (50, 190).

Further, recent studies (132) have revealed the presence in melanocytes of a melanosomal protein different from tyrosinase, which has the ability to catalyze the rearrangement of dopachrome to DHICA. This enzymic reaction is highly stereospecific for normal L-dopachrome, is unaffected by metal chelators and has an optimal pH of about 6.8. Different names have been proposed for this enzyme, *i.e.* dopachrome conversion factor (132, 256), dopachrome oxidoreductase (143), dopachrome isomerase (201), and dopachrome tautomerase (4). It is of interest that another enzyme named dopaquinoneimine conversion factor seems to exist which has the remarkable ability to catalyze the decarboxylative rearrangement of dopachrome to DHI rather than DHICA (193).

According to Pawelek *et al.* (200), the biosynthesis of melanin in Cloudman melanoma cells is a complex process and is regulated by three factors: (a) a dopamine conversion factor which converts dopamine to 5,6-dihydroxyindole (**13**), (b) a 5,6-dihydroxyindole conversion factor which catalyzes the conversion of 5,6-dihydroxyindole to melanin and is active when cells are exposed to melanotropin (MSH), and (c) a 5,6-dihydroxyindole blocking factor which restricts melanogenesis at the 5,6-dihydroxyindole stage. They have also shown that at least three steps in the Raper-Mason scheme of melanin formation from tyrosine are catalysed by tyrosinase (Fig. 6).

4.2.1.2. Polymerization of DHI

To understand the chemistry involved in formation of eumelanin studies dealing with the polymerization of DHI leading to melanochrome (first detected by Mason, 1948), are of great importance. According to Beer *et al.* (7), the broad absorption spectrum at 540–560 nm exhibited by melanochrome was more consistent with a DHI-dimer or oligomer, and they suggested that the 3- and 7-positions were involved in the oxidative coupling of DHI (30). Later, Swan (258) and Kirby *et al.* (128) showed on the basis of tracer studies that during melanin formation each of the various positions of the side chain and the benzene ring of dopa were involved to a similar extent.

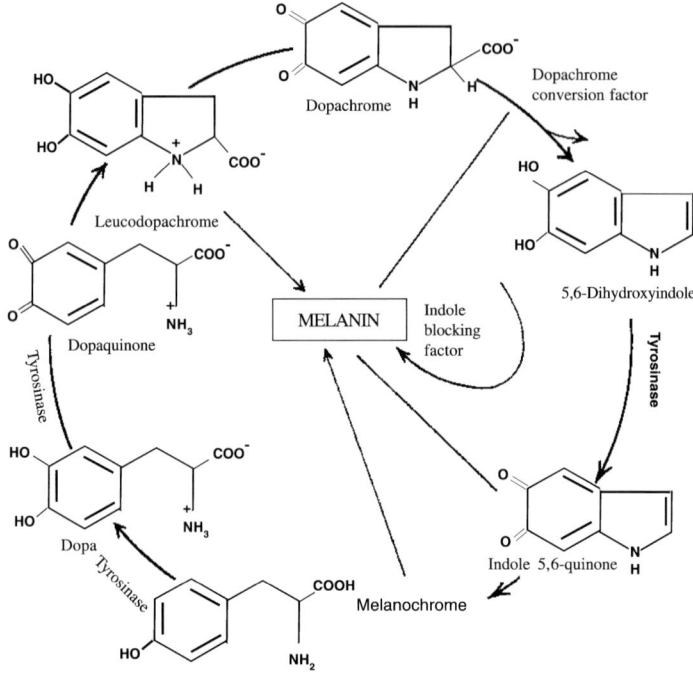

Fig. 6. Regulation of melanogenesis

Much later, Prota *et al.* (*78*), using improved procedures, isolated from a mixture of oligomers at the melanochrome stage three 2,2-, 2,4- and 2,7-dimers and three related trimers. They concluded that the dominant mode of coupling of DHI involved the 2- and 4-positions of the indole ring, with a minor contribution of C-7.

Kinetic experiments and pulse irradiation studies on the mechanism of DHI polymerization (*2, 136*) revealed that the dominating species formed by DHI oxidation was the quinone methide and that coupling proceeds *via* oxygen-centered semiquinone radicals, which can also account for the complexity of the later stages of melanogenesis and the heterogeneity of melanin structure.

The oxidation of DHICA was also studied in the absence and presence of DHI, and two dimers (*191*) and one mixed dimer (*180*) were identified. The overall results obtained in the oxidation of DHI or DHICA at the melanochrome stage support the concept of eumelanin as an intimate mixture of homopolymers of DHI and DHICA, and copolymers of the two indole units in different proportions, the latter depending upon the ratio of formation of DHI and DHICA in the rearrangement of dopachrome.

References, pp. 171–185

4.2.2. Non-Enzymatic Melanin Synthesis: Model Reaction

It appears from the previous discussion that melanogenesis *in vivo* or *in vitro* is regulated by various factors. Hence, to study the dynamics of melanin formation (monooxygenase reaction, *i.e.* constructive metabolism) and breakdown (dioxygenase reaction, *i.e.* catabolism of melanin precursors), a non-enzymatic melanin synthesis from tyrosine and tryptophan, respectively, was devised by Roy *et al.* (*227*), using a prototype of a monooxygenase reaction, *i.e.* the Udenfriend reaction (Fe^{+2}/EDTA/ ascorbic acid) (*272*).

4.2.2.1. The Udenfriend System: A Model for Mixed Function Oxidase

A truly enormous number of different enzymes has been characterized as mixed function oxidases (MFOs) by Mason (*162*) and as monooxygenases by Hayaishi (*107*). They can catalyze a wide variety of oxidation reactions including hydroxylation of aromatic and aliphatic compounds, epoxidation of olefins, oxidative decarboxylation, lactonization *etc.* (*104*). According to Hamilton (*105*), MFO-catalyzed reactions proceed *via* an "oxene mechanism" and there is now ample evidence that the "oxinoid species" is a reduced form of the enzyme (or an enzyme with a reduced cofactor, *e.g.* metal ions) which reacts with oxygen and the substrate either in one step or more than one (*273*).

Many MFO-catalyzed reactions possess characteristic similarities to model reactions and since these enzymatic reactions occur very fast and the intermediates are too short lived, non-enzymatic model systems have been developed to study oxene mechanism. One specific example is the Udenfriend system (*272*) which involves the hydroxylation of aromatic compounds under physiological conditions by a mixture of ascorbate, Fe(II), EDTA, and O_2. According to Hamilton (*105*), the intermediate complex **A** formed during the oxidation reaction (Fig. 7), is believed to be an oxinoid species which transfers electrons to the oxygen atom in the transition state and probably, at the same time, allows the transfer of singlet oxygen from complex **A** to substrate in one step so that diradical intermediates would not be necessary (*105*).

The possible function of an EDTA ligand is to protect the nucleophilic oxygen in **A** from intermolecular reactions although the system operates also without it, even though more slowly (*272, 273*).

In many cases two metal ions can perform the function of the reducing agent; when oxygen is transferred to the substrate each metal ion is oxidized by one electron. This seems to be the explanation for many

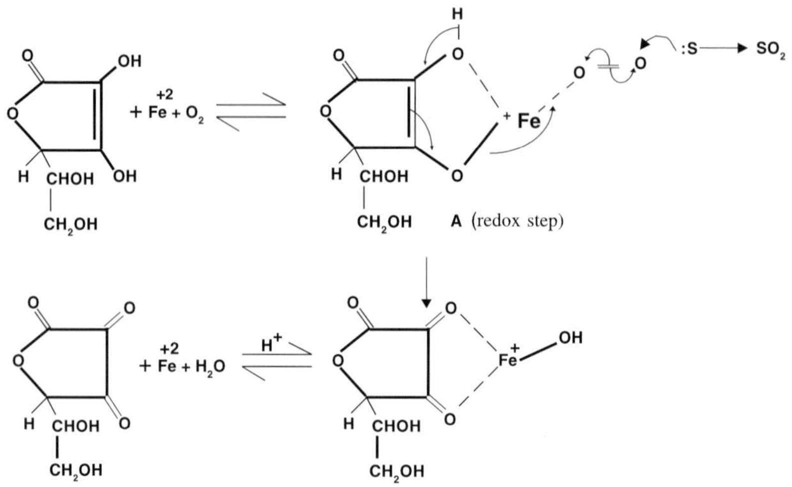

Fig. 7. Mechanism for the oxidation by Udenfriend system

model oxidation reactions as well as for enzyme-catalyzed oxidation reactions involving reduced metal ions (96, 273). Hence, a similar mechanism can be written for the non-specific copper tyrosinase, in which the orthodiphenol intermediate takes the place of ascorbate (274, 275). In fact the mechanism shown in Fig. 3 (see page 152) is very similar to the one in Fig. 7.

4.2.2.2. Melanin Formation Under Udenfriend Conditions

The biomimetic synthesis of melanin from indole accompanied by other biproducts using the Udenfriend reaction was first reported in 1981 (227). Since the obligatory formation of an indole skeleton during the synthesis of melanin from tyrosine may also arise from tryptophan with loss of the latter's three carbon side chain, Chakraborty et al. (51) proposed to study the participation of tryptophan in melanogenesis under Udenfriend conditions. For comparison, experiments with tyrosine as a substrate were also carried out. Thus, in an atmosphere of oxygen, a mixture of ascorbic acid, EDTA, $FeSO_4$, $7H_2O$, and L-tryptophan (L-tyrosine), in $0.1\,M$ phosphate buffer, pH 6.7, was stirred for 240 min with tryptophan and 120 min with tyrosine, respectively. The black particles deposited on standing were collected by centrifugation, the aqueous solutions were extracted with ether, and the ether extracts were worked up by the usual procedures. The yield of the isolated constituents was, however, very low although reproducible.

References, pp. 171–185

Both tryptophan and tyrosine furnished melanin. Similar results were also obtained by Allegri et al. (*3*), who studied melanin synthesis from tryptophan and tyrosine spectrophotometrically. From tyrosine, *p*-hydroxyphenylpyruvic acid (**15**), 4,4-dihydroxybiphenyl (**16**), 5,6-dihydroxyindole (**17**), and 3,5,6-trihydroxyindole (**18**) were obtained in addition to melanin and from tryptophan, 5,6-dihydroxyindole (**17**), indole (**19**), anthranilic acid (**20**), 3-hydroxyanthranilic acid (**21**), indolylpyruvic acid (**22**), 3-hydroxypyrrol-4,5-dicarboxylic acid (**23**), and isatin (**24**).

The results of this biomimetic melanin synthesis are significant. For example, isatin (**24**) as a monoamino oxidase inhibitor may aid higher catecholamine concentration favouring depigmentation (*1*). Oxidation products **20**, **21**, **23** formed from tryptophan could arise through a reaction resembling the *in vivo* dioxygenase reaction on the intermediates leading to 5,6-dihydroxyindole and melanin (*50*). Compounds **18** and **23** could be considered to be candidates for participation in detoxication mechanisms of **17** as indole itself is detoxicated through its 3-hydroxy derivative (*110*). Such a chemical transformation of 5,6-dihydroxyindole might relieve the cell from the cytotoxic effect of **17** (*110*).

Further, owing to deamination, *p*-hydroxyphenylpyruvic acid (**15**), a well-known inhibitor of tyrosinase, was produced from tyrosine. This product underwent further transformation to yield biphenyl **16**; while indolylpyruvic acid (**22**) and subsequently indole (**19**) were formed from tryptophan. Hence, in biological systems, the loss of the starting material, *i.e.* tyrosine, for melanin synthesis due to the formation of **15**, **16** and **18** may be counteracted by tryptophan, as an alternative substrate in the pathway of melanogenesis through **22** and **19**. However, such a replacement of substrate for melanin by tryptophan is not possible in subjects with higher tryptophan pyrrolase (TP) activity, as occurs in vitiligo (*51*).

The overall results of the model reactions suggest that *in vitro* melanin synthesis is a complex process requiring the regulation of various participating reactions (*i.e.* mono-oxygenase/dioxygenase/deamination). Probably the formation of the end products is regulated by the kinetics of the reactions. The results may have relevance to the factors regulating melanin formation *in vivo* according to Pawelek (*200*).

5. Vitiligo

5.1. Introduction

Vitiligo refers to an idiopathic, usually progressive, cosmetic disfigurement of skin due to depigmentation that starts after birth and is not fatal (*188, 189*). The earliest information regarding the disease is in the Ebers Papyrus (*90*) according to which vitiligo was treatable. In the sacred Indian volume "Atharva Veda" dating to 1400 B.C. the condition of "Shweta Kustha" (white leprosy) refers to vitiligo (*280*). "Bohak", "bahah", and "baras" are Arabic names (Koran – 3:48, 5:109) and "kilas" is the Buddhist name for vitiligo (624–544 B.C.).

India has the largest population suffering from vitiligo (1.7%) (*188*). The most characteristic histological picture of this disease is the depletion of skin melanin; the epidermis and dermis are otherwise normal. Since there are no specific biochemical features of the disease, rigid laboratory criteria for a diagnosis of vitiligo are lacking. It is uncertain whether vitiligo is one disease entity with a specific pathogenesis or a common expression of several different lesions.

Although, vitiligo is generally considered to be an acquired condition, several cases of "congenital vitiligo" have been described (*122, 147*). A genetic predisposition is considered to be involved in this disease (*72, 188*).

5.2. Melanocytotoxicity: Antimelanocyte-Antibodies Formation

Several hypotheses regarding the etiology of vitiligo have been reported, but none can totally explain the genesis of this disease (*19, 144, 185, 188, 189, 215*). However, there are four prevailing theories of the pathogenesis of vitiligo which are briefly described in the following.

References, pp. 171–185

5.2.1. The Immune Hypothesis

Two immunological mechanisms could explain the pathogenesis of vitiligo (*155*). A primary disturbance in the immunologic system may result in autoimmunization with the formation of autoantibodies against an antigen of the melanogenic system. This causes melanocytotoxicity, or antimelanocyte-antibodies formation. An alternative mechanism is that there is some injury to the melanocytes which results in the release of a toxic antigenic substance so that formation of antimelanocyte-antibodies ensues and melanogenesis is inhibited (*29, 155, 182*).

5.2.2. The Neural Hypothesis

The neural hypothesis suggests that vitiligo results from the accumulation of some neurochemical mediator which reduces melanin formation (*148, 152*). Both melanocytes and nerve cells are of neuroectodermal origin and both utilize tyrosine to produce an end product, melanin or catecholamine. The similarity of the structures of dopa and catechol suggests that a translation error of receptor sites controlling melanogenesis may cause the aberration of vitiligo (*29, 152*).

5.2.3. The Self-Destruction Hypothesis

This hypothesis suggests that there are intermediates or metabolites in the melanin-synthesis pathway which, being unchecked, lead to melanosome destruction, *i.e.* melanocytes' dysfunction or death (*121, 153*).

5.2.4. The Composite Hypothesis

The above three hypotheses are not mutually exclusive. An immunologic event may be secondary to cutaneous injury or neural stimulation may lead to overproduction of the toxic precursors in melanocytes with subsequent leakage of an aggressive immunologic process destructive to melanocytes (*188*).

Alternatively, melanin formation and destruction may be seen as a physiologically precarious balance process with a metastable equilibrium in those genetically disposed towards it. Overstimulation of neural elements, trauma, sunburn etc., may upset this homeostasis in favour of melanin destruction, again with incontinence of antigenic material and resultant immunologic melanocyte destruction (*35, 144, 189*).

Finally, it cannot be excluded that the primary event is a deficit in feedback control from keratinocytes to melanocytes, such that whatever mechanism genetically limits the number and packaging of melanosomes in keratinocytes, it becomes defective and dominant so as to "turn off" melanocytes completely (*123*). This too could implicate neural, immunologic and self-destruction factors.

Although, this composite hypothesis, like the other three hypotheses, lacks experimental support it appears to encompass more of the facts and abnormalities in vitiliginous patients.

5.3. Chemotherapy of Vitiligo

5.3.1. Psoralens

Chemotherapy of vitiligo begins with use of the indigenous drugs *Ammi majus* in Egypt (1948) and *Psoralea corylifolia* in India (1957). El Mofty (*82*) in Egypt successfully used the 8-methoxy derivative of psoralen (**8-MOP**; **25**) against vitiligo which was put on the market under the trade name Meladenin, while in India, Roychowdhury and Chakraborty (*228*) introduced psoralen (**26**) isolated from *Psoralea corylifolia* as a chemotheraputic agent for treatment of vitiligo.

Xanthotoxin (8-MOP) Psoralen Trimethylpsoralen (TMP)
(**25**) (**26**) (**27**)

Although psoralens do not constitute a magic bullet for the therapy of vitiligo, they are still being used as chemotherapeutic agents on a major scale. In Egypt and in America, 8-MOP has been applied topically or orally with UV irradiation, later known as PUVA therapy (*195*). In India, psoralen and 8-MOP have been used without UV irradiation in many cases of vitiligo, though the effect of sunlight after psoralen application has not been ruled out. Psoralen is considered a better therapeutic agent against vitiligo than 8-MOP because its lower toxicity requires lower doses (*46, 238*). Trimethylpsoralen (**27**) has, however, been claimed to be better than either by Sehgal (*238*). The availability of pure psoralen in India and abroad is limited for want of standard plant materials.

5.3.2. Psoralen Action and UV Light

Furocoumarins are photodynamic substances which evoke various biological activities after photochemical activation. Thus, when psoralen is applied topically to mammalian skin and irradiated with light, it causes formation of erythema and pigmentation. According to Pathak *et al.* (*196*) and Musajo *et al.* (*178*), the pigmentation caused by the psoralen group of drugs is due to photoactivation of the chemical species derived from these drugs. ESR studies have indicated that excitation of the psoralen molecule to the triplet state may lead to free radical formation, which most probably causes biological changes in the irradiated system (*196*).

Kinetic studies of the photoreaction between psoralen and DNA have been carried out. Cole (*63*) reported that interstrand cross-links are formed in native DNA by irradiation with 360 nm light in the presence of psoralen, and suggested that the cross-links may result from the reaction of an excited psoralen molecule with pyrimidine bases in opposite strands of the DNA duplex.

5.3.3. Psoralen Action on Melanogenesis

Pigmentation induced by psoralen drugs plus UV irradiation is evidently due to an increase of functioning melanocytes, increase in the number of melanosomes, increase in the activity of the enzyme, and in the transfer of melanin granules to malpighian cells (*196*).

Psoralen regenerates melanin on vitiliginous spots (*228*) and Chakraborty *et al.* (*43*) reported that psoralen, when applied topically and orally to *Bufo melanostictus*, enhanced melanin formation by accelerating the activity of the melanogenic enzyme tyrosinase (*43*). These results are consistent with the reciprocal relationship between melanization and tyrosinase activity in melanosomes, *i.e.* melanin granules, reported by Seiji *et al.* earlier (*240*). Tyrosinase activation with 8-MOP on vitiliginous skin was also reported by Rudowska (*223*). Lerner *et al.* (*146*), however, found that 8-MOP is without any effect on tyrosinase activation in isolated skin sections of *Xenopus leivis*. On toads, psoralen was more effective orally than topically which simulates the results when psoralen is applied to the skin of human vitiliginous subjects. The use of amphibian subjects like *Bufo melanostictus* as an experimental model is logical, based on observation of Burger and Van Oordt (*31*), as well as on the use of these species by other workers in their research on melanogenesis (*97, 159*).

Studies of hypophysectomized todes by Indian workers revealed that psoralen in the absence of light stimulates tyrosinase activity in the skin and liver of toads (*13, 14*). Carter *et al.* (*34*) found a 2- to 3-fold increase in tyrosinase activity as compared to psoralen alone after UV irradiation of Cloudman melanoma containing TMP. Borkovic *et al.* (*20*), on the basis of their experimental results on melanization with TMP, suggested that TMP may be used in a clinical setting without UV radiation for the regeneration of melanin on vitiliginous spots. Stimulation of melanogenesis with the psoralen group of drugs in the absence of UV light was also reported by other workers (*45, 159, 172, 228*).

It is pertinent here to mention that psoralen exhibits some biological functions which do not require photoactivation such as growth inhibitory properties (*245*), estrogenic activity (*6*), antifungal activity (*42*) and antibiotic activity (*44*). All these observations may be cited to support the concept of Chakraborty (*45*) that psoralen augments melanogenesis through deep seated biochemical events, even without UV irradiation, although additive or synergistic effects of photoactivation are not ruled out.

5.4. Abnormal Biochemical Parameters in Vitiligo

Since none of the hypothetical concepts, acquired or congenital, can totally explain the genesis of vitiligo and since it is well known that depletion of skin melanin is the main criterion for diagnosis of the disease, to get a better understanding of the etiology of the disease, a search for

Table 1. Abnormal biochemical parameters in vitiligo

Parameters	Vitiliginous subjects	References
1. Urinary indole profile	Larger excretion of 5-hydroxyindole acetic acid, 5-hydroxy tryptamine, and kynurenine are consistently present but indican is absent	48, *135*, 229
2. Urinary anthranilic acid, 3-hydroxy anthranilic acid	Higher than normal	48, 229
3. Blood- and skin-SH constant	Higher than normal	*135*
4. Skin lead and tin level	Increased	50, 263
5. Skin nickel, cobalt, and copper level	Insignificant change after depigmentation	50
6. Urinary hydroquinone level	Decreased	40, *192*
7. Serum tyrosinase	Slightly increased	*39, 49*
8. Serum tryptophan pyrrolase (TP) and indoleamine-2,3-dioxygenase (IOD) in skin and lever (dioxygenase activity)	Increased	37, *39, 49*
9. Tyrosine aminotransferase (TAT), Tryptophan pyrrolase (TP), and corticosteroids (Parameters of stress conditions)	Increased	52, *148, 203, 205*

References, pp. 171–185

some biochemical abnormalities associated with the depigmentation in vitiligo was carried out. However, no definite biochemical abnormalities can be considered characteristic of vitiligo skin and it seems most likely that these scattered abnormalities are secondary to the primary process in vitiligo. In most cases reversal of these conditions has been observed during regeneration of pigment with psoralen.

According to Chakraborty *et al.* (*229*), these findings (Table 1) indicate the involvement of abnormal tryptophan metabolism in vitiligo. Lerner (*148*) suggested that stress conditions which are associated with higher TP, TAT, and corticosteroid levels in biological systems (*203*) could be a factor in the precipitation of vitiligo. Further, the activities of dioxygenases, like TP, IOD, associated with vitiligo are superoxide anion (a toxic species of oxygen) dependent (*84*). Hence, overproduction of this reactive oxygen species by varieties of cell stresses may be involved to play an etiological role in vitiligo (radical hypothesis) (*22, 71, 165, 219*).

5.5. Status of Tryptophan in the Melanogenic System

The above results together with previous findings suggest that tryptophan could also be an alternative substrate for melanin synthesis. Chen and Chavin (*60*) used tryptophan as a substrate for goldfish tyrosinase while the incorporation of tryptophan in the melanin of Harding Passey mouse melanoma cells was reported by other authors (*67*). Tryptophan in a true melanin synthesizing system, *i.e.* in B16–F10 melanoma cells cultured in the presence of $2\,mM$ tryptophan, increased melanin formation on the second day (*41*).

The conventional substrates, tyrosine and dopa, can stimulate tyrosinase activity as well as melanization of the cells (*166, 206, 248, 249*). Positive regulation of tyrosinase by its precursors *in vivo* that require active protein synthesis was shown for the first time by Slominski *et al.* (*246, 247*). Both L-tyrosinase and L-dopa stimulate tyrosinase at the level of translation (*248*). Depending on dose and time, L-dopa can both stimulate and inhibit tyrosinase mRNA expression (*249*). Recently, it has been found that tryptophan can stimulate new mRNA-dependent tyrosinase synthesis in B16 murine melanoma cells (*38*).

It has also been observed that tryptophan, like dopa, inhibits tyrosine hydroxylase and dopa oxidase activity of melanosomal tyrosinase and that its inhibitory mechanism differs from inhibition caused by non-substrate type compounds like cysteine and ascorbic acid (*36*). In fact, tyrosinase is inhibited by its own substrate *in vitro* and this inhibition mechanism differs from that caused by cysteine and ascorbic acid (*242, 268*).

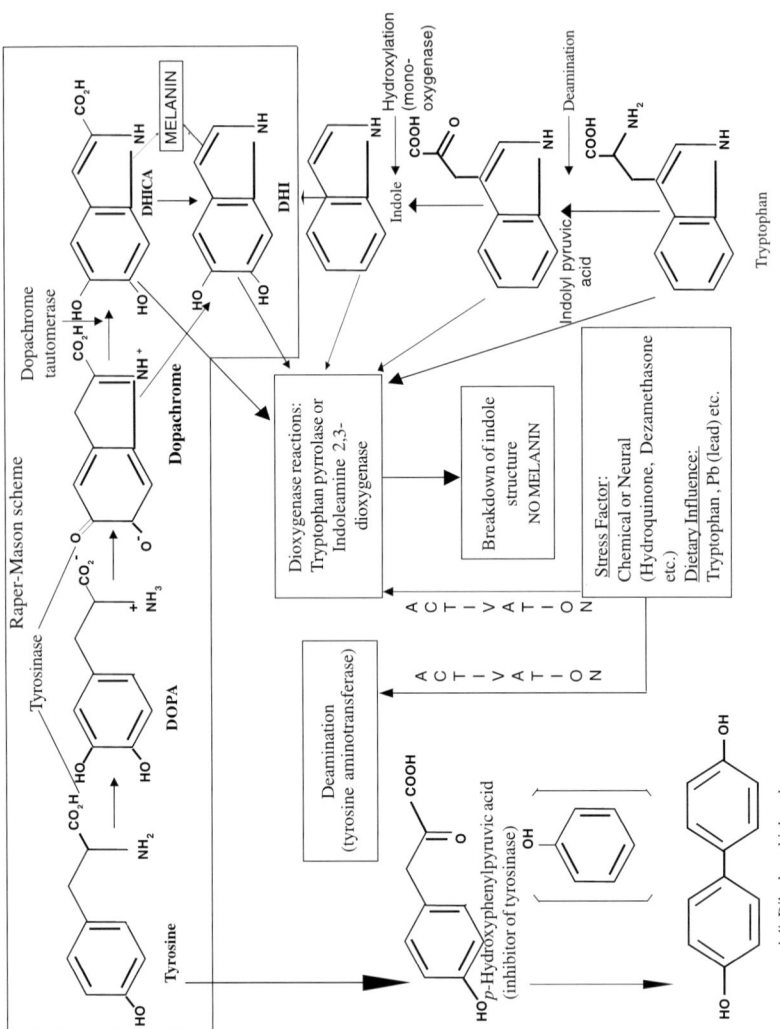

Fig. 8. Combined tyrosine and tryptophan participation for the synthesis of melanin and its catabolism in relation to vitiligo (modification of the Raper-Mason scheme)

All these results show the close resemblance of tryptophan to tyrosine and dopa, with respect to their role as melanin precursors as well as positive regulators of tyrosinase.

5.6. A Composite Hypothesis on Vitiligo

Because both tryptophan and tyrosine participate in melanogenic processes, the possible influence of different enzymes like tyrosinase, tryptophan pyrrolase (TP), indoleamine 2,3-dioxygenase (IOD), and tyrosine aminotransferase (TAT) on the regulation of melanin synthesis as it relates to vitiligo has been depicted in Fig. 8, which constitutes a modification of the Raper-Mason scheme for melanin synthesis (53, 54).

Also in view of the results discussed earlier, it seems that inhibitors like hydroquinones, dexamethasone may increase the amount of catabolic enzymes associated with stress conditions as well as the *in vivo* release of phenyl radicals, facilitating depigmentation (15, 220). Stress not only releases inhibitors of tyrosinase, *e.g.* catecholamines, but also induces TAT *in vivo* (1) causing deamination of tyrosine and yielding *p*-hydroxyphenylpyruvic acid, which is an inhibitor of tyrosinase. Stress has also a great effect on the immune system (127) and on brain-immune system interaction as well (179).

Therefore, the conjecture that foreign inhibitory substances introduced through diet (including nutritional factors) or otherwise may have a role in precipitating and promoting conditions favoring vitiligo may have some validity. The experimental results further suggest that autocytotoxic substances and also superoxide-mediated cell damage (22) may be factors in the impairment of melanogenic processes in vitiligo by immunologic or neurologic mediation (29), thus supporting the composite hypothesis (144, 189) for the pathogenesis of vitiligo.

References

1. Alexandrova M (1994) Stress Induced Tyrosine Aminotransferase Activity *via* Glucocorticoid Receptor. Hormone Metabolic Res **26**: 97
2. Al-Kazwini AT, O'Neill P, Cundall RB, Adams GE, Junino A, Maignan J (1992) Direct Observation of the Reaction of the Quinone Methide from 5,6-Dihydroxyindole with the Nucleophilic Azide Ion. Tetrahedron Lett **33**: 3045
3. Allegri G, Biasiolo M, Frison G, Pelli B, Traldi P (1987) Collisional Spectroscopy in Structural Characterization of Melanins. 1. A First Study on $C_8H_7ON^+$ Ions Originating from Pyrolysis of Biosynthetic and Synthetic Tryptophan Melanins. Pigment Cell Res **1**: 87

4. Aroca P, Garcia-Borron JC, Solano F, Lozano JA (1990) Regulation of Distal Mammalian Melanogenesis. 1: Partial Purification and Characterization of a Dopachrome Converting Factor: Dopachrome Tautomerase. Biochim Biophys Acta **1035**: 266
5. Ban TA, Lehmann HE, Gallai Z, Warnes H, Lee H (1965) Relation Between Photosensitivity and the Pathological Pigmentation of the Skin Produced by High Doses of Chlorpromazine. Union Med Canada **94**: 305
6. Barua A, Banik UK, Chakraborty DP (1961) Endocrinological Studies on Plant Products: Part 1. Preliminary Investigation of Hormonal Action of Psorelen. Ann Biochem Expt Med **21**: 132
7. Beer RJS, Broadhurst T, Robertson A (1954) The Chemistry of the Melanins, Part V: The Autooxidation of 5,6-Dihydroxyindoles. J Chem Soc (London): p. 1947
8. Benathan M, Wyler H (1980) Contribution a l'Analyse Quantitative des Melanines. Yale J Biol Med **53**: 389
9. Bertrand G (1986) Sur un Nouvelle Oxydase on Ferment Soluble Oxydant, d'Origine Vegitable. Comp Rend Acad Sci (Paris) **122**: 1215
10. Billingham RE, Silvers WK (1960) The Melanocytes of Mammals. Q Rev Biol **35**: 1
11. Bischitz PG, Snell RS (1959) The Effect of Testosterone on the Melanocytes and Melanin in the Skin of the Intact and Orchidectomised Male Guinea-Pig. J Inves Dermatol **33**: 299
12. Bischitz PG, Snell RS (1960) A Study of the Effect of Ovariectomy, Estrogen and Progesterone on the Melanocytes and Melanin in the Skin of the Female Guinea-Pig. J Endocrinol **20**: 312
13. Biswas NM, Chakraborty DP, Dev C (1965) Influences of Pituitary on the Melanin Formation in Psoralen Treated Tode (*Bufo melanostictus*). Naturwiss **22**: 622
14. Biswas NM, Dev C, Chakraborty DP (1967) Effects of Psoralen on Pigmentation in Hypophysectomized Tode (*Bufo melanostictus*). Endokrinologie **52**: 271
15. Bleehen SS, Pathak MA, Hori Y, Fitzpatrick TB (1965) Depigmentation of Skin with 4-Isopropylcatechol, Mercaptoamines and Other Compounds. J Invest Dermatol **50**: 103
16. Blois MS (1969) Biological Free Radicals and the Melanins. In: Wyard SJ (ed.) Solid State Biophysics, p. 243. McGraw-Hill, New York
17. Blois MS (1972) The Binding Properties of Melanin: *in vivo* and *in vitro*. In: Montagna W, Stoughton RB, Van Scott EJ (eds.) Advances in Biology of Skin, Vol. XII. Pharmacology and the Skin, p. 65. Appleton-Century-Crofts, New York
18. Blois MS (1978) The Melanins: Their Synthesis and Structure. Photochem Photobiol Rev **3**: 115
19. Bolognia JL, Powelek JM (1988) Biology of Hypopigmentation. Am Acad Dermatol **19**: 217
20. Borkovic SP, Alper JC, McDonald CJ (1983) Stimulation of Pigmentation in Melanoma Cells by Trimethylpsoralen in Absence of Ultraviolet Irradiation. Br J Dermatol **108**: 525
21. Bowness JM, Morton RA (1953) The Association of Zinc and Other Metals with Melanin and Melanin-Protein Complex. Biochem J **53**: 620
22. Bowers RR, Lujan J, Bibso A, Kridel S, Varkey C (1994) Premature Avian Melanocyte Death Due to Low Antioxidant Levels of Protection: Fowl Model for Vitiligo. Pigment Cell Res **7**: 409
23. Boyland E, Willium DC (1956) The Metabolism of Tryptophan. Biochem J **64**: 578

24. Breathnach AS, Bor S, Wyllie LMA (1966) Electron Microscopy of Peripheral Nerve Terminals and Marginal Melanocytes in Vitiligo. J Invest Dermatol **47**: 125
25. Bridelli M, Capelleti R, Crippa PR (1981) Electret State and Hydrated Structure of Melanin. Bioelectrochem Bioeng **8**: 555
26. Brown KT (1965) An Early Potential Evoked by Light from the Pigment Epithelium-Choroid Complex of the Eye of the Tode. Nature **207**: 1249
27. Brown FC, Ward DN, Griffin AC (1959) Preparation and Properties of Mammalian Tyrosinase. In: Gorden M (ed.) Pigment Cell Biology, p. 525. Academic Press, New York
28. Bruenger FW, Stover BJ, Atherton DR (1967) The Incorporation of Various Metal Ions into *in vivo-* and *in vitro-*Produced Melanin. Rad Res **32**: 1
29. Burn R (1972) A Propose de Etiologie du Vitiligo. Dermatologica **145**: 169
30. Bu'Lock JD, Harley-Mason J (1951) Melanin and Its Precursors, Part II: Model Experiments on the Reactions Between Quinones and Indoles, and Consideration of a Possible Structure for the Melanin Polymer. J Chem Soc (London): p. 703
31. Burger AC, Van Oordt GJ (1962) Regulation of Pigment Migration in the Amphibian Melanophore. Gen Comp Endocrinol **1**: 99
32. Cabanes J, Garcia-Canovas F, Lozano JA, Garcia-Carmona P (1987) A Kinetic Study of the Melanization Pathway Between L-Tyrosine and Dopachrome. Biochim Biophys Acta **923**: 187
33. Cassidy HG, Kun KA (1965) In: Oxidation-Reduction Polymer: Redox Polymer. Wiley (Interscience), New York
34. Carter DM, Powelek JM, Candit ES, Koch NG (1974) Stimulation of Tyrosinase by Trimethylpsoralen and Ultraviolet Light in Pigment Cells (Abstr). J Invest Dermatol **62**: 347
35. de Castro LF (1958) Neuropigmentary System and Cell Division in the Maintenance of the Vital Equilibrium of Tissue; A New Theory. Hospital (Rio de Janeiro) **53**: 553
36. Chakraborty AK, Chakraborty DP (1993) The Effect of Tryptophan on Dopa-Oxidation on Melanosomal Tyrosinase. Int J Biochem **25**: 1277
37. Chakraborty AK, Chakraborty A, Chakraborty DP (1993) Hydroquinone Simultaneously Induces Indoleamine-2,3-dioxygenase and Inhibits Tyrosinase in *Bufo melanostictus*. Life Sci **52**: 1695
38. Chakraborty AK, Pawelek J: Unpublished data
39. Chakraborty C, Chakraborty AK, Dutta AK, Chakraborty DP (1983) Abnormal Tryptophan Pyrrolase and Amino Acids Related to Melanogenesis in Vitiligo. Experientia **39**: 280
40. Chakraborty C, Chatterjee A, Chakraborty AK, Chakraborty DP (1984) Inverse-Relationship Between Melanogenesis and Endogenous Hydroquinone. Experientia **40**: 829
41. Chakraborty C, Ichihashi M, Ueda M, Mishima Y, Chakraborty DP (1986) Effects of Tryptophan on Melanogenesis in B16-F10 Melanoma Cells in Culture. IRCS Med Sci **14**: 463
42. Chakraborty DP, Das Gupta A, Bose PK (1957) On the Antifungal Activity of Some Natural Coumarins. Ann Biochem Expt Med **17**: 59
43. Chakraborty DP, Dev C, Mukherjee M (1959) Effect of Psoralen on Tyrosinase Activity and Melanin Pigmentation. Sci Cult **25**: 386
44. Chakraborty DP, Sen M, Bose PK (1961) On the Antibiotic Activity of Some Natural Coumarins. Trans Bose Res Inst **24**: 31
45. Chakraborty DP (1966) Some Biochemical and Chemotherapeutic Aspects of Vitiligo. Indian J Dermatol **11**: 1

46. Chakraborty DP (1968) Some Aspects of Psoralen Drugs. Sci Cult **34**: 33
47. Chakraborty DP, Chatterjee A, Chakraborty AK (1977) In: Abstracts. Indian Chemical Society Symposium. Jaipur, India
48. Chakraborty DP, Chatterjee A (1977) Urinary Indole Profile of *Bufo melanostictus* During Hydroquinone Induced Leucoderma and Its Regeneration. Clin Chem Acta **79**: 399
49. Chakraborty DP, Roychowdhury SK, Dey RN, Chatterjee A (1978) Interrelationship of Tryptophan Pyrrolase with Tyrosinase in Melanogenesis of *Bufo melanostictus*. Clin Chem Acta **82**: 55
50. Chakraborty DP, Chakraborty C, Ganguly M, Chakraborty AK (1983) Trace Metals and Melanogenesis. Experientia **39**: 282
51. Chakraborty DP, Roy S, Chakraborty AK, Rakshit R (1986) Tryptophan Participation in Melanogenesis: A Biomimetic Study. IRCS Med Sci **14**: 940
52. Chakraborty DP, Roy S, Chakraborty AK, Rakshit R, Chakraborty A (1988) Profile of Stress Conditions During Induced Depigmentation: Tryptophan Participation in Melanogenesis and Composite Hypothesis of Vitiligo. IRCS Med Sci **16**: 21
53. Chakraborty DP, Roy S, Chakraborty AK, Rakshit R (1989) Tryptophan Participation in Melanogenesis: Modification of Raper-Mason-Powelek Scheme of Melanin Formation. J Ind Chem Soc **66**: 699
54. Chakraborty DP, Roy S, Chakraborty AK (1996) Vitiligo, Psoralen, and Melanogenesis: Some Observations and Understanding. Pigment Cell Res **9**: 107
55. Chakraborty DP, Roy S (2003) Chemical and Biological Aspects of Melanin. In: Cordell Geoffreya A (ed.) The Alkaloids, Vol. 60, p. 345
56. Chavin H, Schlesinger W (1967) Effects of Melanin Depigmental Agents Upon Normal Pigment Cells, Melanoma and Tyrosinase Activity. In: Montagna W, Hu H (eds.) Advances in the Biology of Skin, Vol. 1, p. 421. Pergamon Press, New York
57. Chedekel MR, Post PW, Deibel RM, Kalus M (1977) Photodestruction of Phaeomelanin. Photochem Photobiol **26**: 651
58. Chedekel MR, Land EJ, Sinelair RS, Tait D, Truscott TG (1980) Photochemistry of 4-Hydroxybenzothiazole: A Model for Phaeomelanin Photodegradation. J Am Chem Soc **102**: 6587
59. Chedekel MR, Subbarao KV, Bahan P, Schultz TM (1987) Biosynthetic and Structural Studies on Phaeomelanin. Biochim Biophys Acta **912**: 239
59a. Chedekel MR, Patil DG, Rao KV, Murphy BP, Clark M, Gardella J, Schultz TM (1988) Solid Phase Carbon ^{13}NMR of ^{13}C-Enriched Eumelanins: The Fate of the Pyrrolic Ring. Pigment Cell Res **1**: 282
60. Chen YM, Chavin W (1965) Radiometric Assay of Tyrosinase and Theoretical Consideration for Melanin Formation. Analyt Biochem **13**: 234
61. Chen YM, Chavin W (1972) Effects of Depigmentary Agents and Related Compounds Upon *in vitro* Tyrosinase Activity. In: Riley V (ed.) Pigmentation, Its Genesis and Biological Control, p. 593. Appleton-Century-Crofts, New York
62. Chio SS, Hyde JS, Sealy RC (1980) Temperature Dependent Paramagnetism in Melanin Polymers. Arch Biochem Biophys **199**: 133
63. Cole RS (1970) Light-Induced Cross-Linking of DNA in the Presence of Furocoumarin. Biochim Biophys Acta **217**: 30
64. Commoner B, Townsend J, Pake GE (1954) Free Radicals in Biological Materials. Nature **174**: 689
65. Cope FW, Sever RJ, Polis BD (1963) Reversible Free Radical Generation in the Melanin Granules of the Eye by Visible Light. Arch Biochem Biophys **100**: 171

66. Costantini C, Crescenzi O, Prota G (1991) Mechanism of the Rearrangement of Dopachrome to 5,6-Dihydroxyindole. Tetrahedron Lett **31**: 3849
67. Costa C, Allegri G, DeAntoni A (1975) Studies on Melanogenesis of Tryptophan in Harding-Passey Mouse Melanoma. Acta Vitaminol Enzymol **29**: 223
68. Courrier R, Cehovic G (1960) Action of Purified Melanophorotropic Hormone (α-MSH) on Thyroid Function in the Rabbit. Compt Rend Acad Sci **251**: 832
69. Crippa R, Cristofoletti V, Romeo N (1978) A Band Model for Melanin Deduced from Optical Absorption and Photoconductivity Experiments. Biochim Biophys Acta **538**: 164
70. Crippa R, Horok V, Prota G, Svoronos P, Wolfram L (1989) Chemistry of Melanins. In: Brossi A (ed.) The Alkaloids, Vol. 36, p. 253. Academic Press, New York
71. Cross CC, Halliwell B, Borrish ET, Pryor WA, Ames BN, Saul RL, McCord JM, Harman D (1987) Oxygen Radicals and Human Disease. Ann Intern Med **107**: 526
72. a) Das SK, Majumdar P, Chakraborty R, Majumdar TK, Halder B (1985) Studies in Vitiligo, 1. Epidemiological Profile in Calcutta (India). Gene Epidemiol **2**: 71; b) Das SK, Majumdar PP, Majumdar TK, Halder B (1985) Studies in Vitiligo, II. Familial Aggregation and Genetics. Gene Epidemiol **2**: 255
73. Delacretaz J (1965) Effect of Testosterone on Tyrosinase Activity on an Experimental Melanoma S-91 of Mice. Ann Dermatol Syph **92**: 25
74. de Mol NJ, Beijersbergen van Henegouwen GMJ, Gerritsma KW (1979) Photochemical Decomposition of Catecholamines–II. The Extent of Aminochrome Formation from Adrenaline, Isoprenaline and Noradrenaline Induced by Ultraviolet Light. Photochem Photobiol **29**: 7479
75. d'Ischia M, Palumbo A, Prota G (1985) 5,6-Dihydroxyindole-2-carboxylic Acid by Treatment of Sepiomelanin with Sodium Borohydride. Tetrahedron Lett **26**: 2801
76. a) d'Ischia M, Napolitano A, Prota G (1987) Sulphydril Compounds in Melanogenesis, Part I: Reaction of Cysteine and Glutathion with 5,6-Dihydroxyindoles. Tetrahedron **43**: 5351; b) d'Ischia M, Napolitano A, Prota G (1987) Sulphydril Compounds in Melanogenesis, Part II: Reaction of Cysteine and Glutathion with Dopachrome. Tetrahedron **43**: 5357
77. d'Ischia M, Prota G (1987) Photooxidation of 5,6-Dihydroxy-1-methyl-indole. Tetrahedron **43**: 431
78. d'Ischia M, Napolitano A, Tsiakas K, Prota G (1990) New Intermediates in the Oxidative Polymerization of 5,6-Dihydroxyindole to Melanin Promoted by Peroxidase/H_2O_2 System. Tetrahedron **46**: 5789 and refs. therein
79. Drochmans P (1961) Study by Electron Microscope of the Mechanism of Melanin Pigmentation; the Distribution of Melanin Granules in the Malpighian Cells. Pathol Biol **9**: 947
79a. Dworzanski JP (1983) J Anal Appl Pytolysis **5**: 69; Chem Abstr **99**: 35447x (1983)
80. Dyster-Aas K, Krakau CET (1964) Increased Permeability of the Blood-Aqueous Humor Barrier in the Rabbit's Eye Provoked by Melanocyte-Stimulating Peptides. Endocrinology **74**: 255
81. Ebrey TG, Cone RA (1967) Melanin, a Possible Pigment for the Photostable Electrical Responses of the Eye. Nature **213**: 360
82. El Mofty AM (1948) A Preliminary Clinical Report on the Treatment of Leucoderma with *Ammi majus* Linn. J Egypt Med Assoc **31**: 651
83. Erway L, Hurley L, Fraser A (1966) Neurological Defect: Manganese in Phenocopy and Prevention of a Genetic Abnormality of Inner Ear. Science **152**: 1766
84. Feigelson P, Brady FO (1974) In: Hayaishi O (ed.) Molecular Mechanism of Oxygen Activation, p. 87. Academic Press, New York

85. Felix CC, Hyde JS, Sarna T, Sealy RC (1978) Interaction of Melanin with Metal Ions. Electron Spin Resonance Evidence for Chelate Complexes of Metal Ions with Free Radicals. J Am Chem Soc **100**: 3922
86. Fitzpatrick TB, Becker Jr SW, Lerner AB, Montgomery H (1950) Tyrosinase in Human Skin: Demonstration of Its Presence and Its Role in Human Melanin Formation. Science **112**: 223
87. Fitzpatrick TB (1952) Human Melanogenesis. Arch Dermatol Syph **65**: 379
88. Fitzpatrick TB, Szabo G (1959) The Melanocytes: Cytology and Cytochemistry. J Invest Dermat **32**: 197
89. Fitzpatrick TB, Miyamoto M, Ishikawa K (1967) The Evolution of Concepts of Melanin Biology. Arch Dermatol **96**: 305
90. Fitzpatrick TB, Mihm MC (1971) In: Fitzpatrick TB, Eisen HS, Wolff K, Freedberg M, Austen KF (eds.) Dermatology in General Medicine, p. 1596. McGraw Hill, New York
91. Flesch P (1949) Inhibitory Action of Extracts of Mammalian Skin on Pigment Formation. Proc Soc Exp Biol NY **70**: 136
92. Foster M (1959) Physiological Genetic Aspects of Mammalian Melanogenesis. Ann NY Acad Sci **100**: 743
93. Fox DL, Kuchnow KP (1965) Reversible Light-Screening Pigment of Elasmobranch Eyes: Chemical Identity with Melanin. Science **150**: 612
94. Fox DL (1976) In: Animal Biochromes and Structural Color, 2nd edn. University of California Press, Berkeley
95. Frenk E (1975) Pigment Cell Biology and Its Relation to Disorders of Melanin Pigmentation (Review). Dermatologica **159**: 185
96. Friedman S, Kaufman S (1965) 3,4-Dihydroxyphenylethylamine β-Hydroxylase. Physical Properties, Copper Content, and Role of Copper in the Catalytic Activity. J Biol Chem **240**: 4763
97. Fukuzawa T, Bagnara JT (1989) Control of Melanoblast Differentiation in Amphiba by α-Melanocyte Stimulating Hormone, a Serum Melanization Factor, and a Melanization Inhibiting Factor. Pigment Cell Res **2**: 171
98. a) Galvao DS, Caldas MJ (1990) Theoretical Investigation of Model Polymers for Eumelanins, I: Finite and Infinite Polymers. J Chem Phys **92**: 2630; b) Galvao DS, Caldas MJ (1990) Theoretical Investigation of Model Polymers for Eumelanins, II: Isolated Defects. J Chem Phys **93**: 2848
99. Gan EV, Haverman HF, Menon IA (1976) Electron Transfer Properties of Melanin. Arch Biochem Biophys **173**: 666
100. Gilchrest BA, Blog FB, Szabo G (1979) Effects of Aging and Chronic Sun Exposure on Melanocytes in Human Skin. J Invest Dermatol **73**: 141
101. Hadley E, Quevedo W (1967) The Role of Epidermal Melanocytes in Adaptive Color Changes in Amphibians. In: Montagna W, Hu F (eds.) Advances in Biology of Skin, Vol. VIII. Pergmon Press, Oxford
102. Hall TC, McCracken BH, Thorn GW (1953) Skin Pigmentation in Relation to Adrenal Cortical Function. J Clin Endocrinol Metabol **13**: 243
103. Hamilton JB (1939) Significance of Sex Hormones in Tanning of the Skin of Women. Proc Soc Exptl Biol Med **40**: 502
104. Hamilton GA, Workman RJ, Woo L (1964) Oxidation by Molecular Oxygen. I. Reactions of a Possible Model System for Mixed-Function Oxidases. J Am Chem Soc **86**: 3390
105. Hamilton GA (1964) Oxidation by Molecular Oxygen. II. The Oxygen Atom Transfer Mechanism for Mixed-Function Oxidases and the Model for Mixed-Function Oxidases. J Am Chem Soc **86**: 3391

106. Hawley MD, Tatawawadi SV, Piekarski S, Adams RN (1967) Electrochemical Studies of the Oxidation Pathways of Catecholamines. J Am Chem Soc **89**: 447
107. Hayaishi O (1964) Proc 6th Intern Congr Biochem, Plenary Sessions, New York **33**: 31
108. Hearing VJ, Ekel TM, Montague PM, Nicholson JM (1980) Mammalian Tyrosinase. Stoichiometry and Measurement of Reaction Products. Biochim Biophys Acta **611**: 251
109. Hempel K (1966) Investigation on the Structure of Melanin in Malignant Melanoma with ^{3}H- and ^{14}C-Dopa Labelled at Different Positions. In: Della Prota G, Muhlbock O (eds.) Structure and Control of Melanocytes, p. 162. Springer, Berlin Heidelberg New York
110. Hochstein P, Cohen G (1963) The Cytotoxicity of Melanin Precursors. Ann NY Acad Sci **100**: 876
111. Holstein TJ, Stowell CP, Quevedo Jr WC, Zarearo RM, Bienieki TC (1973) Peroxidase, "Protyronase" and the Multiple Forms of Tyrosinase in Mice. Yale J Biol Med **46**: 560
112. Horok V, Gillette JR (1971) A Study of the Oxidation-Reduction State of Synthetic 3,4-Dihydroxy-DL-phenylalanin Melanin. Molec Pharmacol **7**: 429
113. Horok V: Unpublished experimental observation
114. Hunter JAA, Mottaz JH, Zelickson AS (1970) Melanogenesis: Ultrastructural Histochemical Observations on Ultraviolet Irradiated Human Melanocytes. J Invest Dermatol **54**: 213
115. Imokawa G, Yada Y, Hori Y (1987) In: Jimbow K (ed.) The Structure and Function of Melanin, Vol. 4, p. 60. Fuji-Shoin Co. Ltd., Sapporo, Japan
116. Ippen H (1961) The Pineal Body Hormone Melatonin and the Central Pigment Regulation. Deut Med Wochenschr **86**: 307
117. Ito S, Prota G (1977) A Facile One-Step Synthesis of Cysteinyldopa Using Mushroom Tyrosinase. Experientia **33**: 1118
118. Ito S, Palumbo A, Prota G (1985) Tyrosinase-Catalyzed Conjugation of Dopa with Glutathion. Experientia **41**: 960
119. Ito S (1986) Reexamination of the Structure of Eumelanin. Biochim Biophys Acta **883**: 155
120. Ito S, Imai Y, Kato T, Fujita K (1987) In: Jimbow K (ed.) Structure and Function of Melanin, p. 44. Fuji-Shoin Co. Ltd., Sapporo, Japan
121. Jimbow K, Szabo G, Fitzpatrick TB (1974) Ultrastructural Investigation of Autophagocytosis of Melanosomes and Programmed Death of Melanocytes in White Leghorn Feathers: A Study of Morphogenetic Events Leading to Hypomelanosis. Dev Biol **36**: 8
122. Jimbow K, Fitzpatrick TB, Szabo G, Hori Y (1975) Congenital Circumscribed Hypomelanosis: A Characterization Based on Electronmicroscopic Study of Tuberous Sclerosis, *Nevus depigmentosus*, and Piebaldism. J Invest Dermatol **64**: 50
123. Jimbow K, Quevedo WC, Fitzpatrick TB, Szabo G (1976) Some Aspects of Melanin Biology (Review): 1950–1975. J Invest Dermatol **67**: 72
124. Kalyanaraman B, Felix CC, Sealy RC (1985) Semiquinone Anion Radicals of Catecholamines, Catechol, Estrogens, and Their Metal Ion Complexes. Environ Health Perspect **64**: 185
125. Kean EA (1964) A Procedure Which Demonstrates Substrate Inhibition of Tyrosinase. Biochem Biophys Acta **92**: 602
126. Kertesz D (1954) The Phenol Oxidizing Enzyme System of Human Melanomas: Substrate Specificity and Relationship to Copper. J Nat Cancer Inst **14**: 1081

127. Khansari DN, Murgo AJ, Faith RE (1990) Effects of Stress on the Immune System (Review). Immunol Today **11**: 170
128. Kirby GW, Ogunkoya L (1965) Structure of Melanin Derived from ($\pm$)-3,4-Dihydroxy-[^{14}C, ^{3}H]-phenylalanin by Oxidation with Tyrosinase. Chem Commun **21**: 546
129. Kissinger PT, Hineman WR (1984) In: Laboratory Techniques in Electroanalytical Chemistry, p. 611. Dekker, New York
130. Kono R, Yamaoka H, McGinness J (1979) Anomalous Absorption and Dispersion of Sound Wave in Diethylamine Melanin. J App Phys **50**: 1242
131. Kosto B, Pickford GE, Foster M (1959) Further Studies of the Hormonal Induction of Melanogenesis in Killifish, *Fundulus heteroclitus*. Endocrinol **65**: 869
132. Korner A, Powelek J (1980) Dopachrome Conversion: A Possible Control Point in Melanin Biosynthesis. J Invest Dermatol **75**: 192
133. Korytowski W, Kalyanaraman B, Menon IA, Sarna T, Sealy RC (1986) Reaction of Superoxide Anions with Melanins: Electron Spin Resonance and Spin Trapping Studies. Biochim Biophys Acta **882**: 145
134. Krivoy WA, Guillemin R (1961) On a Possible Role of β-Melanocyte-Stimulating Hormone (β-MSH) in the Central Nervous System of the Mammalians; an Effect of β-MSH in the Spinal Cord of the Cat. Endocrinol **68**: 170
135. Kurbanov KH, Berezov TT (1976) Tryptophan Metabolism in Vitiligo (Russian). Vopr Med Khim **22**: 683
136. Lambert C, Chacon JN, Chedekel MR, Land EJ, Riley PA, Thompson A, Truscott G (1989) A Pulse Radiolysis Investigation of the Oxidation of Indolic Melanin Precursors: Evidence for Indolequinones and Subsequent Intermediates. Biochim Biophys Acta **993**: 12
137. Lancy ME (1984) Phonon-Electron Coupling as a Possible Transducing Mechanism in Bioelectronic Process Involving Neuromelanin. J Theor Biol **111**: 201
138. Lapina VA, Dontsov AE, Ostrovskii MA (1984) a) Chem Abstr (1985) **102**: 2110j; b) Chem Abstr (1984) **101**: 187528x
139. Larsson B, Tjalve H (1978) Studies on the Melanin Affinity of Metal Ions. Acta Physiol Scand **104**: 479
140. Larsson B, Oskarsson A, Tjalve H (1978) On the Binding of the Bisquaternary Ammonium Compound Paraquat to Melanin and Cartilage *in vivo*. Biochem Pharmacol **27**: 1721
141. Larsson B, Tjalve H (1979) Studies on the Mechanism of Drug-Binding to Melanin. Biochem Pharmacol **28**: 1181
142. Lee TH, Lerner AB (1959) Melanocyte-Stimulating Hormones from Pituitary Glands. In: Gordon M (ed.) Pigment Cell Biology, p. 435. Academic Press, New York
143. Leonard LJ, Townsend D, King RA (1988) Dopachrome Oxidoreductase and Metal Ions in Dopachrome Conversion in the Eumelanin Pathway. Biochemistry **27**: 6156
144. LePoole IC, Das P, Van den Wijngaard R, Bos J, Westerhof W (1993) Review of the Etiopathomechanism of Vitiligo: A Convergence Theory. Exp Dermatol **2**: 145
145. Lerner AB, Fitzpatrick TB (1950) Biochemistry of Melanin Formation. Physiol Rev **30**: 91
146. Lerner AB, Denton CR, Fitzpatrick TB (1953) Clinical and Experimental Studies with 8-Methoxypsoralen in Vitiligo. J Invest Dermatol **20**: 299
147. Lerner AB, Lerner MR (1958) Congenital and Hereditary Disturbances of Pigmentation. Bibliotheca paediat **66**: 308

148. Lerner AB (1959) Vitiligo. J Invest Dermatol **32**: 285
149. Lerner AB, Wright MR (1960) *In vitro* Frog Skin Assay for Agents That Darken and Lighten Melanocytes. Methods of Biochem Anal **8**: 295
150. Lerner AB, McGuire JS (1961) Effect of Alpha- and Beta-Melanocyte Stimulating Hormones on the Skin Color of Man. Nature **189**: 176
151. Lerner AB, McGuire JS (1964) Melanocyte-Stimulating Hormone and Adrenocorticotropic Hormone. Their Relation to Pigmentation. New Engl J Med **270**: 535
152. Lerner AB (1971) Neural Control of Pigment Cells. In: Kawamura T, Fitzpatrick TB, Seiji M (eds.) Biology of Normal and Abnormal Melanocytes, p. 3. Tokyo University Press, Tokyo
153. Lerner AB (1971) On the Etiology of Vitiligo and Gray Hair. Am J Med **51**: 141
154. Long JM, Krivoy WA, Guillemin R (1961) On a Possible Role of β-Melanocyte-Stimulating Hormone (β-MSH) in the Central Nervous System of Mammalians; Enzymatic Inactivation *in vitro* of β-MSH by Brain Tissue. Endocrinol **68**: 176
155. Lorinez AL (1954) In: Rothman S (ed.) Pigmentation in Physiology and Biochemistry of the Skin, p. 515. University of Chicago Press, Chicago
156. Lukiewicz S (1972) The Biological Role of Melanin. 1. New Concepts and Methodical Approaches. Fol Histochem Cytochem **10**: 93
157. Malmstrom BG, Lars R (1968) Copper Containing Oxidases. In: Singer TP (ed.) Biological Oxidation, p. 415. Academic Press, New York, London
158. Marsden CD (1969) Brain Melanin. In: Wolman M (ed.) Pigments in Pathology, p. 395. Academic Press, New York
159. Marwan MM, Jiang JW, de Lauro Castrucci AM, Hadley ME (1990) Psoralen Stimulate Mouse Melanocyte and Melanoma Tyrosinase Activity in the Absence of Ultraviolet Light. Pigment Cell Res **3**: 214
160. Mason HS (1955) Comparative Biochemistry of the Phenolase Complex. Adv Enzymol **16**: 105
161. Mason HS, Fowlks WL, Peterson E (1955) Oxygen Transfer and Electron Transport by the Phenolase Complex. J Am Chem Soc **77**: 2914
162. Mason HS (1957) Mechanisms of Oxygen Metabolism. Adv Enzymol **19**: 79
163. Mason HS (1967) The Structure of Melanin. In: Montagna W, Hu F (eds.) Advances in Biology of Skin, Vol. VIII, p. 293. Pergamon Press, Oxford
164. Mayausky JS, McCreery RL (1983) Spectroelectrochemical Examination of Charge Transfer Between Chlorpromazine Cation Radical and Catecholamines. Ann Chem **55**: 308
165. McCord JM (1987) Oxygen-Derived Radicals: A Link Between Reperfusion Injury and Inflammation. Fed Proc **46**: 2402
166. McEwan M, Parsons PG (1987) Inhibition of Melanization in Human Melanoma Cells by a Serotonin Uptake Inhibitor. J Invest Dermatol **89**: 82
167. McGinness J (1972) A Mechanism for Band Gaps in Melanins. Science **177**: 896
168. McGinness J, Proctor P (1973) The Importance of the Fact That Melanin is Black. J Theor Biol **39**: 677
169. McGinness J, Corry P, Proctor P (1974) Amorphous Semiconductor Switching in Melanins. Science **183**: 853
170. McGinness J (1985) A New View of Pigmented Neurons. J Theor Biol **115**: 475
171. McGuire J (1966) Melanin Granule Dispersion in Epidermal Melanocytes. J Invest Dermatol **45**: 547; and refs therein

172. Mengeaud V, Ortonne JP (1994) Regulation of Melanogenesis Induced by 5-Methoxypsoralen Without Ultraviolet Light in Murine Melanoma Cells. Pigment Cell Res **7**: 245
173. Mizutani U, Massalski TB, McGinness JE, Corry PM (1976) Low Temperature Specific Heat Anomalies in Melanins and Tumor Melanosomes. Nature (London) **259**: 505
174. Mohn MP (1957) Inhibitory Effects of Adrenaline on Skin, Hair Growth, and Pigmentation in Black Rats. Anat Record **127**: 337
175. Montagna W, Porta G, Kenney J (1993) In: Black Skin. Academic Press, New York
176. Mott NF (1967) Electrons in Disordered Structures. Adv Phys **16**: 49
177. Mottaz JH, Zelickson AS (1967) Melanin Transfer: A Possible Phagocytic Process. J Invest Dermatol **49**: 605
178. Musajo L, Rodighiero G (1962) The Skin Photosensitizing Furocoumarins. Experientia **18**: 153
179. Nagata S (1993) Stress-Induced Immune Changes, and Brain–Immune Interaction (Review) (Japanese). Sangy Ika Daigaku Zasshi **15**: 161
180. Napolitano A, Crescenzi O, Prota G (1993) Copolymerization of 5,6-Dihydroxyindole and 5,6-Dihydroxyindole-2-carboxylic Acid in Melanogenesis. Isolation of a Cross-Coupling Product. Tetrahedron Lett **34**: 885
181. Nathanson L (1967) Biological Aspects of Human Malignant Melanoma. Cancer **20**: 650
182. Naughton GK (1983) Detection of Antibodies to Melanocytes in Vitiligo by Specific Immunoprecipitation. J Invest Dermatol **81**: 540
183. Nelson RM, Mason HS (1970) Tyrosinase (Mushroom). In: Tabor H, Tabor CW (eds.) Methods in Enzymology XVII A, p. 626. Academic Press, New York, London
184. Nicolaus RA (1968) Melanins. Hermann, Paris
185. Nordlund JJ, Ortonee JP (1992) Vitiligo and Depigmentation. Curr Prob Dermatol **4**: 3
186. Oikawa A, Nakayasu M (1973) Yale J Biol Med **46**: 500; Chem Abstr **81**: 74292r (1974)
187. Okun MR, Edelstein LM, Hamada G, Bulmental G, Donnellan B, Burnett J (1972) Oxidation of Tyrosine and Dopa to Melanin by Mammalian Peroxidase: The Possible Role of Peroxidase in Melanin Synthesis and Catecholamine Synthesis *in vivo*. In: Riley V (ed.) Pigmentation, Its Genesis and Biological Control, p. 571. Appleton-Century-Crofts, New York
188. Ortonee JP, Mosher DB, Fitzpatrick TB (1983) Approach to the Problem of Leucoderma. In: Ortonee JP, Mosher DB, Fitzpatrick TB (eds.) Vitiligo and Other Hypomelanosis of Hair and Skin, p. 37. Plenum, New York
189. Ortonee JP, Bose S (1993) Vitiligo: Where Do We Stand? Pigment Cell Res **6**: 61
190. Palumbo A, d'Ischia M, Misuraca G, Prota G (1987) Effect of Metal Ions on the Rearrangement of Dopachrome. Biochim Biophys Acta **925**: 203
191. Palumbo A, d'Ischia M, Prota G (1987) Tyrosinase Promoted Oxidation of 5,6-Dihydroxyindole-2-carboxylic Acid to Melanin. Isolation and Characterization of Oligomer Intermediates. Tetrahedron **43**: 4203
192. Palumbo A, d'Ischia M, Misuraca M, Prota G (1992) Skin Pigmentation by Hydroquinone: A Chemical and Biochemical Insight. Pigment Cell Res **2**: 299
193. Palumbo A, d'Ischia M, Misuraca G, De Martino L, Prota G (1994) A New Dopachrome Rearranging Enzyme from the Ejected Ink of the Cuttlefish *Sepia officinalis*. Biochem J **299**: 839
194. Parker GH (1948) Animal Color Changes and Their Neurohumours. Cambridge University Press, London

195. Parrish TA, Fitzpatrick TB, Shea C, Pathak MA (1976) Photochemotherapy of Vitiligo: Use of Orally Administrated Psoralens and a High-Intensity Long-Wave Ultraviolet Light System. Arch Dermatol **112**: 1531
196. Pathak MA, Allen B, Ingram DJE, Fellman JH (1961) Photosensitization and the Effect of Ultraviolet Radiation on the Production of Unpaired Electron in Presence of Furocoumarins (Psoralen). Biochem Biophys Acta **54**: 506
197. Pathak MA, Riley FC, Fitzpatrick TB (1962) Melanogenesis in Human Skin Following Exposure to Long-Wave Ultraviolet and Visible Light. J Invest Dermatol **39**: 435
198. Pathak MA, Fitzpatrick TB (1974) In: Fitzpatrick TB, Pathak MA, Harber LC, Seiji M, Kukita A (eds.) Sunlight and Man, p. 725. University of Tokyo, Tokyo
199. Patil DG, Chedekel MR (1984) Synthesis and Analysis of Pheomelanin Degradation Products. J Org Chem **49**: 997; and refs. therein
200. Pawelek J, Korner A, Bergstrom A, Bologna J (1980) New Regulators of Melanin Biosynthesis and the Autodestruction of Melanoma Cells. Nature **286**: 617; and refs. therein
201. Pawelek JM (1990) Dopachrome Conversion Factor Functions as an Isomerase. Biochim Biophys Res Commun **166**: 1328
202. Pawelek J (1991) After Dopachrome? Pigment Cell Res **4**: 53
203. Peranio C, Lamar Jr C, Pitot HC (1966) Studies on the Mechanism of Carbohydrate Repression in Rat Liver (Review). Adv Enzym Regul **4**: 199
204. Piattelli M, Fattorusso E, Magno S, Nicolaus RA (1963) The Structure of Melanins and Melanogenesis, III: The Structure of Sepiomelanin. Tetrahedron **19**: 2061; and refs. therein
205. Pittner RA, Fears R, Brindley DN (1985) Effects of cAMP, Glucocorticoids and Insulin on the Activities of Phosphatidate Phosphohydrolase, Tyrosine Aminotransferase and Glycerol Kinase in Isolated Rat Hepatocytes in Relation to the Control of Triglycerol Synthesis and Gluconeogenesis. Biochem J **225**: 455
206. Price JE, Tarin D, Fidler IJ (1988) Influence of Organ Microenvironment on Pigmentation of a Metastatic Murine Melanoma. Cancer Res **48**: 2258
207. Proctor P, McGinness J, Corry P (1974) A Hypothesis on the Preferential Destruction of Melanized Tissues. J Theor Biol **48**: 19
208. Prota G (1988) In: Bagnata H (ed.) Advances in Pigment Cell Research, p. 101. Academic Press, New York
209. Prota G, d'Ischia M, Napolitano A (1988) The Regulatory Role of Sulphydryl Compounds in Melanogenesis. Pigment Cell Res **1S**: 48
210. Prota G (1992) Melanins and Melanogenesis. Academic Press, San Diego
211. Prota G (1995) The Chemistry of Melanins and Melanogenesis In: Herz W, Falk H, Kirby GW (eds.) Fortschr Chem Organ Naturstoffe, Vol. 64, p. 93. Springer, Wien New York
212. Pullman A, Pullman B (1961) The Band Structure of Melanins. Biochim Biophys Acta **54**: 384
213. Quevedo Jr WC, Szabo G, Birks J, Sinesi SJ (1965) Melanocyte Populations in UV-Radiated Human Skin. J Invest Dermatol **45**: 295
214. Quevedo WC, Fitzpatrick TB, Pathak MA, Jimbow K (1974) Light and Skin Colour. In: Pathak MA, Herber LC, Seiji M, Kukita A, Fitzpatrick TB (eds.) Sunlight and Man, p. 165. Tokyo University Press, Tokyo
215. Ramaiah A (1994) Vitiligo. J Ind Chem Soc **71**: 355
216. Raper HS (1928) The Aerobic Oxidases. Physiol Rev **8**: 245

217. Raper HS (1938) Some Problems of Tyrosine Metabolism. J Chem Soc: p. 125
218. Remers WA (1972) Properties and Reactions of Indoles. In: Houlihan WI (ed.) Indoles, Part I, p. 152. Wiley, New York
219. Repine JE, Pfenninger OW, Talmage DW, Berger EM, Pettijohn DE (1981) Dimethyl Sulfoxide Prevents DNA Nicking Mediated by Ionizing Radiation or Iron/Hydrogen Peroxide-Generated Hydroxyl Radicals. Proc Natl Acad Sci USA **78**: 1001
220. Riley PA (1970) Mechanism of Pigment Cell Toxicity Produced by Hydroxyanisol. J Pathol **101**: 163
221. Robb DA (1984) Tyrosinase. In: Loutie R (ed.) Copper Proteins, Vol. 2, p. 207. Florida CRC Press, Boca Raton
222. Robson NC, Swan GA (1966) Studies on the Structure of Some Synthetic Melanins. In: Della Prota G, Muhlbock O (eds.) Structure and Control of the Melanocytes, p. 155. Springer, Berlin Heidelberg New York
223. Rudowska I (1965) Studies on the Mechanism of Psoralens in the Treatment of Vitiligo (Polish). Prezeglad Dermatologiczny **52**: 391
224. Rodriguez-Lopez JN, Tudela J, Varon R, Garcia-Canovas F (1991) Kinetic Study on the Effect of pH on the Melanin Biosynthesis Pathway. Biochim Biophys Acta **1076**: 379
225. Rorsman H, Albertson E, Edholm LE, Hansson C, Ogren L, Rosengren E (1988) Thiols in the Melanocyte. Pigment Cell Res **1S**: 54
225a. Rosel MA, Mosca L, DeMarco C (1992) Melanin Production from Enkephalins by Tyrosinase. Biochem Biophys Res Commun **184**: 1190
226. Ross GT, Odell WD (1963) The Effects of Temperature on the Response to Melanocytes-Stimulating Hormone. Ann NY Acad Sci **100**: 696
227. Roy S, Chakraborty AK, Chakraborty DP (1981) Melanin Formation and Breakdown of Indole Under Udenfriend Conditions. J Ind Chem Soc **58**: 992
228. Roychowdhury Jr A, Chakraborty DP (1957) Psoralen Therapy in Leucoderma. Cal Med J **54**: 139
229. Roychowdhury SK, Chakraborty DP (1968) Some Urinary Indole Profile of Vitiliginous Patients. Clin Chim Acta **22**: 298
230. Sarna T, Froncisz W, Hyde JS (1980) Cu^{2+} Probe of Metal-Ion Binding Sites in Melanin Using Electron Paramagnetic Resonance Spectroscopy. Natural Melanin II. Arch Biochem Biophys **202**: 304; and refs. therein
231. Sarna T, Duleba A, Korytowski W, Swartz H (1980) Interaction of Melanin with Oxygen. Arch Biochem Biophys **200**: 140
232. Sarna T, Menon IA, Sealy RC (1984) Photoinduced Oxygen Consumption in Melanin Systems – II. Action Spectra and Quantum Yields for Phaeomelanins. Photochem Photobiol **39**: 805
233. Sarna T, Korytowski W, Sealy RC (1985) Nitroxides as Redox Probe of Melanins: Dark-Induced and Photo-Induced Changes in Redox Equilibria. Arch Biochem Biophys **239**: 266
234. Sarna T (1992) Properties and Function of the Ocular Melanin: A Photobiophysical View. J Photochem Photobiol **B12**: 215
235. Schally AV, Kastin AJ, Redding TW, Bowers CY, Yaijma H, Kubo K (1967) Thyroid Stimulating and Pigmentary Effects of Synthetic Peptides Related to α-MSH and ACTH. Metabolism **16**: 824
236. Sealy RC, Felix CC, Hyde JS, Swartz HM (1980) Structure and Reactivity of Melanins: Influence of Free Radicals and Metal Ions. In: Pryor WA (ed.) Free Radicals in Biology, Vol. 4, p. 209. Academic Press, New York

237. Sealy RC, Sarna T, Wanner EJ, Reszka K (1984) Photosensitization of Melanin: An Electron Spin Resonance Study of Sensitized Radical Production and Oxygen Consumption. Photochem Photobiol **40**: 453
238. Sehgal VN (1975) A Comparative Clinical Evolution of Trimethylpsoralen, Psoralen, and 8-Methoxypsoralen in Treating Vitiligo. Int J Dermatol **14**: 205
239. Seiji M, Fitzpatrick TB, Birbeck MS (1961) The Melanosome: A Distinctive Subcellular Particle of Mammalian Melanocytes and the Site of Melanogenesis. J Invest Dermatol **36**: 243
240. Seiji M, Fitzpatrick TB (1961) The Reciprocal Relationship Between Melanization and Tyrosinase Activity in Melanosomes (Melanin Granules). J Biochem **49**: 700
241. Seiji M, Fitzpatrick TB, Simpson RT, Birbeck MSC (1963) Chemical Composition and Terminology of Specialized Organelles (Melanosomes and Melanin Granules) in Mammalian Melanocytes. Nature **197**: 1082
242. Seiji M, Iwashita S (1966) Intracellular Localization of Tyrosinase and Site of Melanin Formation in Melanocyte. J Invest Dermatol **45**: 305
243. Seiji M, Sasaki M, Tomita Y (1978) Nature of Tyrosinase Inactivation in Melanosomes. Tohoku J Exp Med **125**: 233
244. Shibata T, Prota G, Mishima Y (1993) Non-Melanosomal Regulatory Factors in Melanogenesis. J Invest Dermatol **100**: 274S
245. Sinha Roy SP, Chakraborty DP (1976) Psoralen, a Powerful Germination Inhibitor. Phytochemistry **15**: 1205
246. Slominski A, Moellmann G, Kuklinska E, Bomirski A, Pawelek J (1988) Positive Regulation of Melanin Pigmentation by Two Key Substrates of the Melanogenic Pathway, L-Tyrosine and L-Dopa. J Cell Sci **89**: 287
247. Slominski A (1989) L-Tyrosine Induces Synthesis of Melanogenesis-Related Proteins. Life Sci **45**: 1799
248. Slominski A, Costantino R (1991) L-Tyrosine Induces Tyrosinase Expression *via* a Posttranscriptional Mechanism. Experientia **47**: 721
249. Slominski A, Costantino R (1991) Molecular Mechanism of Tyrosinase Regulation by L-Dopa in Hamster Melanoma Cells. Life Sci **48**: 2075
250. Snell RS, Bischitz PG (1960) The Effect of Large Doses of Estrogen and Progesterone on Melanin Pigmentation. J Invest Dermatol **35**: 73
251. Snell RS (1961) The Influence of Sex Hormones on Melanin Pigmentation of the Skin. Biochem J **78**: 17
252. Snell RS (1962) Effect of Progesterone on the Activity of Melanocytes in the Skin. Z Zellforsch Mikroskop Anat **57**: 818
253. Snell RS (1965) Effect of Melatonin on Mammalian Epidermal Melanocytes. J Invest Dermatol **44**: 273
254. Staricco RJ, Pinkus H (1957) Quantitative and Qualitative Data on the Pigment Cells of Adult Human Epidermis. J Invest Dermatol **28**: 33
255. Sugumaran M, Semenst V (1990) Formation of a Stable Quinone Methide During Tyrosinase-Catalyzed Oxidation of Alpha-Methyldopa Methyl Ester and Its Implication in Melanin Biosynthesis. Bioorg Chem **18**: 144
256. Sugumaran M (1992) Letter to the Editor. Pigment Cell Res **5**: 203
257. Swan GA, Waggott A (1970) Studies Related to the Chemistry of Melanins. Part X: Quantitative Assessment of Different Types of Units Present in Dopa-Melanin. J Chem Soc (C) (London): p. 1409
258. Swan G (1973) Current Knowledge of Melanin Structure. In: McGovern VJ, Russell P (eds.) Pigment Cell, Vol. 1, p. 151. Karger, Basel

259. Swan GA (1974) Structure, Chemistry and Biosynthesis of the Melanins. In: Herz W, Grisebach H, Kirby GW (eds.) Fortschr Chem Organ Naturstoffe, Vol. 31, p. 521. Springer, Wien New York
260. Szabo G (1967) Tyrosinase in the Epidermal Melanocytes of White Human Skin. Arch Dermatol **76**: 324
261. Takenouchi K (1963) Thiamine and Riboflavine Metabolism in Skin Disease. Toda Printing Company, Chiba City, Japan
262. Takenouchi K, Aso K (1964) The Relation Between Melanin Formation and Ascorbic Acid. J Vitamin (Kyoto) **10**: 123
263. Tenconi LT, Acocella G (1966) Study on the Chemotherapy of Experimental Lead Poisoning. 1. Effects of Lead Poisoning on the Tryptophan to Nicotinic Acid Metabolism in the Rat. Acta Vitaminologica **20**: 189
264. Thomas M (1955) Melanins. In: Peach K, Trecey MV (eds.) Modern Methods of Plant Analysis. Vol. IV, p. 661. Springer, Berlin Göttingen Heidelberg
265. Thomson RH (1974) The Pigments of Raddish Hair and Feathers. Angew Chem Int Ed Engl **13**: 305
266. Thomson RH (1962) Melanins. In: Florkin M, Mason HS (eds.) Comparative Biochemistry, Vol. VIII, Part A, p. 727. Academic Press, New York London
267. Toda K, Hory Y, Fitzpatrick TB (1969) The Site of Tyrosinase Activity Within the Melanosome (Abstr). J Invest Dermatol **52**: 380
268. Tomita Y, Hariu A, Mizuno C, Seiji M (1980) Inactivation of Tyrosinase by Dopa. J Invest Dermatol **75**: 379
269. Tomita Y, Fukushima M, Tagami H (1987) In: Jimbow K (ed.) Structure and Function of Melanin. Vol. 4, p. 14. Fuji Shoin Co. Ltd., Sapporo, Japan
270. Tompsett D (1939) Memoirs on Typical British Marine Plants and Animals, Vol. XXXII. University Press, Liverpool
271. Toshima S, Moore GE, Sandberg AA (1968) Ultrastructure of Human Melanoma in Cell Culture: Electron Microscopic Studies. Cancer **21**: 202
272. Udenfriend S, Clark CT, Axelrod J, Brodie BB (1954) Ascorbic Acid in Aromatic Hydroxylation. 1. A Model System for Aromatic Hydroxylation. J Biol Chem **208**: 731
273. Ullrich V, Staudinger HJ (1966) In: Bloch K, Hayaishi O (eds.) Biological and Chemical Aspects of Oxygenases, p. 235. Maruzen, Tokyo
274. Ullrich V, Duppel W (1975) Iron and Copper Containing Monooxigenases. In: Boyer PD (ed.) The Enzymes, Vol. XII, p. 253. Academic Press, New York London
275. Venneste WH, Zuberbuhler A (1974) In: Hayaishi O (ed.) Molecular Mechanisms of Oxygen Activation, p. 371. Academic Press, New York; and refs. therein
276. Voorhees JJ, Duell EG, Bass LJ, Harrell ER (1973) Role of Cyclic AMP in the Control of Epidermal Cell Growth and Differentiation, Chalones: Concepts and Current Researches. Natal Cancer Inst Monogr **38**: 47; and refs. therein
277. Walaas E (1963) The Chemical Transformation of the Chatecholamines Induced by UV Irradiation. Photochem Photobiol **2**: 9
278. Wheeler MH, Bell AA (1988) Melanins and Their Importance in Pathogenic Fungi. Curr Top Med Mycol **7**: 338
279. Whitney WD (1905) Atharva Veda Samhita (Translation and Notes). Harvard Oriental Series, Vol. 7. Harvard University Press, Lanman, Mass
280. White R, Hu F (1977) Characteristics of Tyrosinase in B 16 Melanoma. J Invest Dermatol **68**: 272
281. Wolfram LJ, Berthiaume M (1985) 6th Eur Workshop Melanin Pigm

282. Woods M, Du Buy H, Burk D (1950) Evidence for the Mitochondrial Nature and Function of Melanin Granules. Zoologica **35**: 30
283. Wurtman RJ, Axelrod J (1966) Effect of Chlorpromazine and Other Drugs on the Disposition of Circulating Melatonin. New Eng J Med: p. 274
284. Yasunobu KT, Peterson EW, Mason HS (1959) The Oxidation of Tyrosine-Containing Peptides by Tyrosinase. J Biol Chem **234**: 3291
285. Young TE, Oriswold JR, Hulbert MH (1974) Melanin. 1: Kinetics of Oxidative Cyclization of Dopa to Dopachrome. J Org Chem **39**: 1980
286. Young TE, Babbitt BW, Wolfe LA (1980) Melanin. 2. Electrochemical Study of the Oxidation of α-Methyldopa and 5,6-Dihydroxy-2-methylindole. J Org Chem **45**: 2899
287. Young TE, Babbitt BW (1983) Electrochemical Study of the Oxidation of α-Methyldopamine, α-Methylnoradrenaline, and Dopamine. J Org Chem **48**: 562
288. Zeise L, Chedekel MR (1992) Melanin Standard Method: Titrimetric Analysis. Pigment Cell Res **5**: 230

Author Index

Page numbers printed in *italics* refer to References

Abbas, H.K.   *117, 120*
Abdusamatov, A.   *57*
Abe, T.   *59*
Aboul-Nasr, M.B.   *119, 122*
Abramson, D.   *118*
Abreu, P.M.   *116*
Achilladelis, B.   *116, 126*
Acocella, G.   *184*
Adams, G.E.   *171*
Adams, P.M.   *114*
Adams, R.N.   *177*
Adams, W.   *115*
Adlercreutz, H.   *118*
Agnello, E.J.   *56*
Ahmed-Schofield, R.   *59*
Aikawa, Y.   *114*
Akai, S.   *60, 61*
Akano, K.   *122, 123*
Aladesanmi, A.J.   *58*
Albertson, E.   *182*
Alexander, N.J.   *128, 129*
Alexandrova, M.   *171*
Al-Hetti, M.B.   *116*
Al-Kazwini, A.T.   *171*
Allegri, G.   163, *171, 175*
Allen, B.   *181*
Allin, S.M.   *58*
Alper, J.C.   *172*
Altamirano, F.   2, 53, *56, 61*
Altomare, C.   *117, 118, 126*
Amer, M.E.   *57*
Ames, B.N.   *175*
Amici, R.M.   *60*
Ammon, H.L.   *119, 124*
Anderson, D.J.   *62*
Anderson, K.S.   *127*
Andrew, I.G.   *117*
Andriamialisoa, R.Z.   *57*
ApSimon, J.W.   *115, 118, 124*
Arigoni, D.   *126*
Aristotle   132, 150

Aroca, P.   *172*
Arrendale, R.F.   *115*
Ashley, A.   *114*
Aso, K.   *184*
Atherton, D.R.   *173*
Avren'eva, L.I.   *118*
Axelrod, J.   *184, 185*
Ayer, W.A.   *114*

Babbitt, B.W.   *185*
Bagnara, J.T.   *176*
Bahan, P.   *174*
Bailey, W.J.   *122, 123*
Baldwin, N.C.P.   *114, 118, 129*
Balta, L.E.   *123*
Bamburg, J.R.   *118, 119*
Ban, T.A.   *172*
Banik, U.K.   *172*
Barnes, C.L.   *115, 119, 125*
Barros, C.S.   *120, 122, 123, 130*
Barros, S.S.   *123*
Bartok, T.   *126*
Barton, D.H.R.   18, 19, *58*
Barua, A.   *172*
Bass, L.J.   *184*
Bassfield, R.L.   *115*
Bata, A.   *123*
Baxter, M.   *117*
Bean, G.A.   *119, 122, 123*
Beavers, W.   *62*
Becker, L.   *62*
Becker Jr, S.W.   *176*
Beer, R.J.S.   159, *172*
Behrens, J.   *115*
Beijersbergen van Henegouwen, G.M.J.   *175*
Bekele, E.   *119, 125*
Bekker, A.R.   *124*
Bell, A.A.   *184*
Belleau, B.   39, *60*
Benathan, M.   136, *172*
Bennett, G.A.   *120*

Beremand, M.N. *115, 124, 125, 127–129*
Berezov, T.T. *178*
Berger, E.M. *182*
Bergers, W.W.A. *116*
Bergstrom, A. *181*
Berthiaume, M. *184*
Bertoni, M.D. *130*
Bertrand, G. *172*
Bhat, R.V. *120*
Bhattacharya, S.K. *62*
Biasiolo, M. *171*
Bibso, A. *172*
Bick, I.R.C. *57*
Bienieki, T.C. *177*
Bilai, V.I. *118*
Billingham, R.E. *172*
Bindseil, K.-U. *116*
Birbeck, M.S. *183*
Birks, J. *181*
Birmingham, J.M. *126*
Bischitz, P.G. *172, 183*
Biswas, K. *117*
Biswas, N.M. *172*
Blackwell, B.A. *114, 115, 117, 118, 124, 126*
Blanchfield, B.J. *117*
Bleehen, S.S. *172*
Blight, M.M. *121*
Bloem, R.J. *120–122*
Blog, F.B. *176*
Blois, M.S. *172*
Boar, R.B. *58*
Boehme, K. *57*
Boekelheide, V. *4, 56, 57*
Boettner, F.E. *123*
Bogdanova, I.A. *118*
Bohm, J. *120*
Bohner, B. *121*
Bojase, G. *57*
Bok, S.H. *114*
Bologna, J. *181*
Bolognia, J.L. *172*
Boltyenskaya, E.V. *115*
Bomirski, A. *183*
Booth, C. *73, 76–78, 96, 113*
Bor, S. *173*
Borkovic, S.P. *167, 172*
Borrish, E.T. *175*
Bos, J. *178*
Bosch, U. *118, 120*

Bose, P.K. *173*
Bose, S. *180*
Bottalico, A. *116, 117, 119, 126*
Böttcher, K. *57, 60*
Bowers, C.Y. *182*
Bowers, R.R. *172*
Bowness, J.M. *172*
Boyland, E. *172*
Brady, F.O. *175*
Braun, H. *60*
Breathnach, A.S. *173*
Breedlove, C.K. *124*
Breitenstein, W. *122, 126*
Brian, P.W. *113*
Bridelli, M. *173*
Bridges, C.H. *123*
Bright, W. *124*
Brindley, D.N. *181*
Brioni, J.D. *62*
Broadhurst, T. *172*
Brodal, G. *121*
Brodie, B.B. *184*
Brown, D.W. *128*
Brown, F.C. *173*
Brown, K.T. *147, 173*
Bruenger, F.W. *173*
Bruton, B. *119*
Bryan, R.F. *124*
Bulmental, G. *180*
Bu'Lock, J.D. *136, 173*
Bunge, R.H. *120, 121, 123*
Bunge, R.W. *122*
Burger, A.C. *167, 173*
Burger, B.V. *121*
Burgess, L.W. *125*
Burk, D. *185*
Burmeister, H.R. *119*
Burn, R. *173*
Burnett, J. *180*
Busam, L. *123, 124*
Bycroft, B.W. *114, 118, 125, 128, 129*

Cabanes, J. *173*
Cabral, D. *130*
Cai, Y.-S. *121*
Caldas, M.J. *176*
Camp, B.J. *123*
Campbell, V. *120*
Candit, E.S. *173*
Cane, D.E. *126, 127*

Capelleti, R. *173*
Carmack, M. *56*
Carter, D.M. 167, *173*
Casper, H.H. 115, *119*, *125*
Cassayre, J. *59*
Cassidy, H.G. *173*
Cassidy, M.P. *59*
Cehovic, G. *175*
Chacon, J.N. *178*
Chakrabarti, D.K. 117, *119–121*
Chakraborti, A. *58*
Chakraborty, A. *173, 174*
Chakraborty, A.K. *173, 174, 182*
Chakraborty, C. *173, 174*
Chakraborty, D.P. 154, 162, 166–169, *172–174, 182, 183*
Chakraborty, R. *175*
Charland, J.-P. *118, 124*
Chatterjee, A. *173, 174*
Chatterjee, K. *121*
Chaudhary, K.C.B. *117, 119*
Chauviere, G. *57*
Chavin, H. *174*
Chavin, W. 169, *174*
Chawla, A.S. *56*
Chawla, H.M. *57*
Chedekel, M.R. 138, *174, 178, 181, 185*
Cheeta, S. *62*
Chelkowski, J. *119*
Chen, Y.M. 169, *174*
Chikaoka, S. *60*
Chio, S.S. *174*
Chiu, H.-T. *127*
Choi, E.J. *114*
Choi, S.U. *114*
Chou, C.T. *59*
Christensen, C.M. *115*
Chu, F.S. *129*
Ciegler, A. *120*
Cigelnik, E. *115*
Clark, C.T. *184*
Clark, J.D. *118*
Clark, M. *174*
Clear, R.M. *118*
Cohen, G. *177*
Cohen, T. *58*
Cole, M. *117*
Cole, P.D. *115*
Cole, R.J. *115, 118*
Cole, R.S. 167, *174*

Combrinck, S. *121*
Comezoglu, F.T. *122*
Comezoglu, S.N. 119, *122–124*
Comezoglu, T.F. *123*
Commoner, B. *174*
Cone, R.A. 147, *175*
Conkova, E. *120*
Cope, F.W. *174*
Corley, D.G. *114–116, 119, 121, 124*
Corry, P. *179–181*
Costa, C. *175*
Costantini, C. *175*
Costantino, R. *183*
Courrier, R. *175*
Cox, C. *120*
Cox, R.H. *115, 118*
Craig, L.E. *62*
Crescenzi, O. *175, 180*
Crippa, P.R. *173*
Crippa, R. *175*
Cristofoletti, V. *175*
Croft, W.A. *122*
Cross, C.C. *175*
Csanyi, E. *116*
Cundall, R.B. *171*
Cunfer, B.M. *115*
Curtis, R.W. *122*
Cutler, H.G. *115, 118*

Dagne, E. *57*
Dailey, R.G. *124*
D'Angelo, J. *61*
Danishefsky, S.J. *59*
Das, P. *178*
Das, S.K. *175*
Das Gupta, A. *173*
Dawkins, A.W. *113*
DeAntoni, A. *175*
Dearborn, D.G. *114, 118*
De Castro, L.F. *173*
Decker, M.W. *62*
Deibel, R.M. *174*
Delacretaz, J. *175*
De Lauro Delaurs Castrucci, A.M. *179*
Del Toro, G.V. *62*
DeMarco, C. *182*
De Martino, L. *180*
De Mol, N.J. *175*
Denton, C.R. *178*
DeSilva, T. *113, 119, 122*

Author Index

Desjardins, A.E. *118, 124, 127–129*
Desmaële, D. *18, 57, 61*
Deur, C. *59*
Deus-Neumann, B. *58*
Dev, C. *172, 173*
Devor, K.A. *115, 124–126, 128*
De Vries, T.J. *62*
Dewick, P.M. *114, 118, 125, 128, 129*
Dey, R.N. *174*
Di Menna, M.E. *114, 120, 122*
D'Ischia, M. *175, 180, 181*
Dmitrieva, I.V. *115*
Dobereiner, J. *124*
Doi, S. *116*
Dominguez, M. *61*
Donnellan, B. *180*
Donnelly-Roberts, D.L. *62*
Dontsov, A.E. *178*
Doohan, F.M. *127*
Dorner, J.W. *115, 118*
Dreifuss, P.A. *123*
Drochmans, P. *175*
Du Buy, H. *185*
Duell, E.G. *184*
Duleba, A. *182*
Dumas, F. *61*
Duppel, W. *184*
Durley, R.C. *119*
Dutta, A.K. *173*
Dutta, S. *62*
Dworzanski, J.P. *175*
Dyke, S.F. *58*
Dyster-Aas, K. *175*

Ebrey, T.G. *147, 175*
Edelstein, L.M. *180*
Edholm, L.E. *182*
Ehrhardt, M. *60*
Ekel, T.M. *177*
El-Banna, A.A. *120*
El Bialy, S.A.A. *60*
Eliel, E.L. *61*
El-Kady, I.A. *114, 118*
Ellanskaya, I.A. *118*
Eller, K.I. *115, 118, 120*
Ellis, J.J. *119*
Ellison, R.A. *118, 120*
El-Maghraby, O.M.O. *118, 119*
El-Masry, S. *57*
El Mofty, A.M. *166, 175*

Elsegood, M.R.J. *58*
Enomoto, M. *117*
Epling, B. *123*
Eppley, R.M. *122, 123*
Erway, L. *175*
Etoh, H. *57*
Ettmayr, C. *57, 61*
Etzel, R.A. *114, 118*
Evans, R. *127, 128*
Evidente, A. *116, 117*

Faith, R.E. *178*
Farah, C.A. *128*
Farinole, F. *119*
Faro, H.P. *57*
Fattorusso, E. *181*
Fears, R. *181*
Feigelson, P. *175*
Fekete, C. *127*
Felix, C.C. *176, 177, 182*
Fellman, J.H. *181*
Fernando, T. *119*
Fetz, E. *121*
Fidler, I.J. *181*
Fielder, D.A. *118, 124*
File, S.E. *62*
Fisher, P.J. *116*
Fitzpatrick, T.B. *157, 172, 176–178, 180, 181, 183, 184*
Fitzsimons, B.C. *127*
Flesch, P. *125, 176*
Flippen-Anderson, J.L. *114, 122, 123*
Flury, F. *117*
Fogliano, V. *118*
Folkers, K. *2, 56*
Forbes, G.A. *125*
Forsyth, G. *123*
Fort, D.M. *119, 125*
Foster, M. *176, 178*
Fowlks, W.L. *179*
Fox, D.L. *176*
Frade, H. *122*
Franck, B. *58*
Franich, R.A. *117*
Fraser, A. *175*
Freckman, W.G. *123*
Freeman, G.G. *121*
French, J.C. *120–123*
Frenk, E. *176*
Frevert, J. *116*

Author Index

Freyer, A.J.   *57*
Friedman, S.   *176*
Frison, G.   *171*
Froncisz, W.   *182*
Fujita, K.   *177*
Fukasaku, K.   *121*
Fukumoto, K.   *56*
Fukushima, M.   *184*
Fukuzawa, T.   *176*
Furneaux, R.H.   *61*

Gainsford, G.J.   *61*
Gallai, Z.   *172*
Galvao, D.S.   *176*
Gams, W.   *125*
Gan, E.V.   *176*
Ganguly, M.   *174*
Garcia-Borron, J.C.   *172*
Garcia-Canovas, F.   *173*, *182*
Garcia-Carmona, P.   *173*
Garcia-Mateos, R.   *61*
Gardella, J.   *174*
Gardner, D.   *116*
Garin-Aguilar, M.E.   *61*, *62*
Gashe, B.A.   *57*
Gauthier, M.J.   *128*
Gelderblom, W.C.A.   *121*
Geoghegan, R.F.   *113*
George, C.F.   *114*, *122*
Gerard, P.J.   *62*
Gerritsma, K.W.   *175*
Gervay, J.E.   *60*
Ghosal, G.   *119*
Ghosal, S.   *58*, *62*, *117*, *120*, *121*
Giczey, G.   *127*
Giese, B.   *59*
Gilardi, R.D.   *114*, *122*
Gilbert, J.   *114*, *118*, *125*, *128*, *129*
Gilchrest, B.A.   *176*
Gilgan, M.W.   *117*
Gillette, J.R.   *177*
Glaz, E.T.   *116*
Gledhill, L.   *125*, *128*
Glen, A.T.   *116*
Gloer, J.B.   *123*
Godtfredsen, W.O.   *113*, *115*
Golinsky, P.   *117*, *120*
Gong, B.   *122*
Gorst-Allman, C.P.   *119*
Goto, S.   *57*

Greenhalgh, R.   *114*, *115*, *117*, *118*, *121*, *124*, *126*
Greshoff, M.   2, *56*
Griffin, A.C.   *173*
Gripenberg, J.   *118*
Grove, J.F.   *113*, *121*
Grunden, M.F.   *56*
Guella, G.   *119*
Guillemin, R.   *178*, *179*
Gunn, D.E.   *60*
Guo, M.   *121*
Gyimesi, J.   *116*

Ha, H.-J.   *126*
Habermehl, G.G.   *116*, *123*, *124*
Hadley, E.   *176*
Hadley, M.E.   *179*
Haidukowski, M.   *122*
Halder, B.   *175*
Hall, T.C.   *176*
Halliwell, B.   *175*
Hamada, G.   *180*
Hamilton, G.A.   *156*, *161*, *176*
Hamilton, J.B.   *176*
Hamilton, P.B.   *116*
Hammerschmidt, R.   *125*
Hamon, J.   *58*
Haney, C.A.   *116*
Hanlin, R.T.   *115*
Hansen, K.F.   *57*
Hansen, U.   *125*
Hanson, J.R.   *114*, *116*, *126–128*
Hansson, C.   *182*
Hao, W.   *122*
Harada, S.   *61*
Hargreaves, R.T.   *62*
Harington, J.S.   *116*
Hariu, A.   *184*
Harley-Mason, J.   *136*, *173*
Harman, D.   *175*
Harrach, B.   *122*, *123*
Harrell, E.R.   *184*
Harri, E.   *114*, *121*, *122*
Hart, J.B.   *62*
Haruna, M.   *60*
Harwig, J.   *117*
Hasselmeyer, G.   *60*
Hauser, D.   *117*
Haverman, H.F.   *176*
Hawley, M.D.   *177*

Hayaishi, O.   161, *177*
Hayashi, M.   *60*
Hearing, V.J.   *177*
Heeg, M.J.   *59*
Hegnauer, M.   *57*
Hegnauer, R.   *57*
Hemming, H.G.   *113*
Hempel, K.   *177*
Hennig, R.   *60*
Hesketh, A.R.   *116, 125, 128*
Heydel, P.   *124*
Highet, R.J.   *122*
Hill, D.W.   *123*
Hineman, W.R.   *178*
Hinkley, S.   *118*
Hintikka, E.-L.   *114, 118, 126*
Hochstein, P.   *177*
Hocking, A.D.   *115*
Hoffmann, J.J.   *58*
Hohn, T.M.   *114, 125, 127–130*
Holden, I.   *118*
Holmlund, C.E.   *113, 119*
Holstein, T.J.   *177*
Hori, Y.   *172, 177*
Horler, H.   *59*
Hornok, L.   *127*
Horok, V.   *175, 177*
Hory, Y.   *184*
Hoshino, Y.   *117*
Hosoi, S.   *58, 59, 61*
Hosoya, T.   *121*
Howell, S.A.   *121*
Hu, F.   *184*
Huang, L.   *125*
Huang, X.   *121*
Hulbert, M.H.   *185*
Hunter, J.A.A.   *177*
Hurley, L.   *175*
Hussain, S.S.   *57*
Hussein, H.M.   *117*
Hyde, J.S.   *174, 176, 182*

Ichihara, A.   *116*
Ichihashi, M.   *173*
Ichinoe, M.   *117, 118, 126*
Iida, A.   *116*
Ikeda, M.   *59–61*
Ilus, T.   *118*
Imai, Y.   *177*
Imokawa, G.   *177*

Ingram, D.J.E.   *181*
Ippen, H.   *177*
Irie, H.   *59*
Isaka, M.   *123*
Isbell, T.A.   *129*
Ishi, Y.   *59*
Ishibashi, H.   *60, 61*
Ishida, K.   *58*
Ishiguro, Y.   *58*
Ishii, K.   *116–118*
Ishikawa, K.   *60, 176*
Isobe, K.   *58*
Ito, K.   *60, 61*
Ito, S.   *177*
Itoh, M.   *116*
Iwashita, S.   *183*

Jakubowski, Z.L.   *123*
James, R.   *58*
James, S.L.   *58*
Jarvis, B.B.   *113, 114, 117–124, 130*
Jenson, A.H.   *120*
Jiang, J.   *114, 118*
Jiang, J.W.   *179*
Jiang, Y.   *115*
Jimbow, K.   *177, 181*
Joffe, A.Z.   *119*
Johns, S.R.   *56*
Johnson, K.   *125*
Johnson, R.D.   *62*
Jong, S.-C.   *126*
Jousse, C.   *57*
Juma, B.F.   *57*
Junino, A.   *171*

Kalus, M.   *174*
Kalyanaraman, B.   *177, 178*
Kametani, T.   *56*
Kanaeda, S.   *117*
Kaneko, C.   *59*
Kaneko, I.   *128*
Kaneko, M.   *58*
Kaneko, Y.   *116*
Kaneuchi, S.   *59, 61*
Kang, C.H.   *62*
Kanhere, S.R.   *114*
Kapoor, V.K.   *56*
Kappe, C.O.   *60*
Kashiwaba, N.   *59*
Kassem, F.F.   *57*

Author Index

Kastin, A.J. *182*
Kasuga, I. *123*
Katagiri, N. *59*
Katayama, K. *116*
Katayama, T. *123*
Kato, T. *177*
Kaufman, S. *176*
Kawamura, O. *121*
Kawasaki, T. *58, 123*
Kawazu, K. *123*
Kayser, O. *116*
Kean, E.A. *177*
Kelley, C.J. *58*
Kemmelmeir, C. *120*
Kenner, G.W. *56*
Kenney, J. *180*
Kertesz, D. *177*
Khansari, D.N. *178*
Khorana, H.G. *56*
Kientz, C.E. *116*
Kim, K.-H. *114, 115*
Kim, N.Y. *114*
Kim, S.U. *114*
Kimura, M. *128, 129*
King, R.A. *178*
Kirby, G.W. *58, 159, 178*
Kishaba, A.N. *122*
Kissinger, P.T. *178*
Kis-Tamas, A. *123*
Kitagawa, T. *59*
Kitahara, T. *58*
Kiuchi, F. *59, 61*
Knox-Davies, P.S. *118*
Kobayashi, A. *123*
Kobayashi, H. *122, 123*
Kobayashi, J. *58*
Koch, N.G. *173*
Komiyama, M. *128*
Kommedahl, T. *117, 123*
Kondo, E. *123*
Konishi, K. *116*
Koniuszy, F. *56*
Kono, R. *178*
Kononenko, G.P. *115, 124*
Korner, A. *178, 181*
Korytowski, W. *178, 182*
Koshino, H. *128*
Kosto, B. *178*
Kotsonis, F.N. *118, 120*
Kowala, C. *56*

Krakau, C.E.T. *175*
Kridel, S. *172*
Kriek, N.P.J. *116*
Krishna, R. *129*
Krivoy, W.A. *178, 179*
Kubo, H. *116*
Kubo, K. *182*
Kuchnow, K.P. *176*
Kuklinska, E. *183*
Kun, K.A. *173*
Kupatashvilli, N. *57*
Kupchan, S.M. *124*
Kurata, H. *114, 118, 126*
Kurbanov, K.H. *178*
Kurisa, M. *117*
Kwon, B.-M. *114*

Lamar Jr, C. *181*
Lambert, C. *178*
Lamberton, J.A. *56*
Lamprecht, S.C. *118*
Lancy, M.E. *178*
Land, E.J. *174, 178*
Langley, P. *116*
Langlois, N. *57, 58*
Langseth, W. *115*
Lansden, J.A. *118*
Lapina, V.A. *178*
Lars, R. *179*
Larsson, B. *178*
Lasztity, R. *123*
Latus-Zietkiewicz, D. *120*
Lau, P.-Y. *120*
Lauren, D.R. *114*
Laurent, D. *119*
Leary, J.D. *58*
Le Dréau, M.A. *61*
Lee, C.O. *114*
Lee, H. *172*
Lee, H.I. *59*
Lee, S.Y. *114*
Lee, T.H. *178*
Lee, Y.-W. *114, 115, 118, 119, 122, 123*
Lehman, A.J. *61*
Lehmann, H.E. *172*
Leonard, L.J. *178*
Leonov, A.L. *124*
Leonov, A.N. *115*
LePoole, I.C. *178*
Lerner, A.B. *167, 169, 176, 178, 179*

Lerner, M.R. *178*
Lertwerawat, Y. *123*
Levandier, D. *115*, *124*
Li, Y. *115*
Liang, P.-H. *127*
Lin, F. *115*
Liu, W. *115*
Livinghouse, T. *61*
Loeffler, W. *114*, *116*
Logrieco, A. *116–118*, *126*, *127*
Long, J.M. *179*
Lorinez, A.L. *179*
Louis-Seize, G. *114*
Loukaci, A. *116*
Lowe, D. *113*
Lozano, J.A. *172*, *173*
Lozoya, M. *61*
Lozoya, X. *61*
Lu, B. *58*
Lu, M. *126*
Lujan, J. *172*
Lukiewicz, S. *179*
Luo, Y. *115*

Ma, G. *126*
Ma, X. *58*
Machida, Y. *114*, *124*, *125*, *128*
MacKenzie, S. *114*
MacKusick, B.C. *56*
Maeda, H. *60*
Magno, S. *181*
Maier, H.U. *58*
Maier, M.S. *130*
Maignan, J. *171*
Majinda, R.T. *57*
Major, R.T. *56*
Majumdar, P. *175*
Majumdar, S.K. *58*
Majumdar, T.K. *175*
Mallet, A.I. *121*
Malloch, J. *123*
Malmstrom, B.G. *179*
Mamer, O. *125*
Mancini, I. *119*
Mandava, N.B. *123*
Mannina, L. *118*
Marasas, W.F.O. *114*, *118*, *121*, *125*
Mariano, P.S. *59*
Marino, J.P. *61*
Marsden, C.D. *179*

Marsh, D.C. *114*, *125*, *128*
Marten, T. *126*, *128*
Martin, W.P. *58*
Martinez-Vasquez, M. *61*
Marwan, M.M. *179*
Mason, H.S. 140, 158, 159, 161, *179*, *180*, *185*
Mason, J.M. *61*, *62*
Massalski, T.B. *180*
Masuma, R. *114*
Matsuda, Y. *118*
Matsui, Y. *121*
Matsumoto, G. *129*
Matsumoto, K. *123*
Matsumoto, M. *123*
Matsuno, K. *59*
Matzenbacher, N.I. *130*
Mauli, R. *116*, *117*
Mayausky, J.S. *179*
Mazzocchi, D.B. *114*
Mazzola, E.P. *113*, *118–120*, *122*, *123*
McAlees, A.J. *115*
McAlpine, J.B. *119*
McCapra, F. *60*
McCord, J.M. *175*, *179*
McCormick, S.P. *114*, *125*, *127–130*
McCracken, B.H. *176*
McCreery, R.L. *179*
McDonald, C.J. *172*
McDonald, E. *61*
McEwan, M. *179*
McGinness, J. 142, 146, *178–181*
McGuire, J. *179*
McGuire, J.S. *179*
McLachlan, A. *115*
Mebs, D. *124*
Meguro, S. *122*, *123*
Meier, R.-M. *115*, *124*
Mengeaud, V. *180*
Menon, I.A. *176*, *178*, *182*
Miao, S. *114*
Midiwo, J.O. *113*, *118*, *119*, *121–123*
Mihm, M.C. *176*
Miller, J.D. *114*, *115*, *117*, *118*, *121*, *124*, *126*
Miller, R.W. *124*
Miller-Wideman, M. *119*
Millington, D.S. *62*
Minato, H. *123*
Mine, Y. *123*
Miranda, L.D. *60*

Mirocha, C.J.   *115–122*
Mishima, Y.   *173, 183*
Misuraca, G.   *180*
Miyamoto, M.   *176*
Mizuno, C.   *184*
Mizutani, U.   *180*
Moellmann, G.   *183*
Mohn, M.P.   *180*
Mohr, P.   *124*
Mokhtari-Rejali, N.   *130*
Mondal, M.H.   *62*
Mondon, A.   18, 39, *57, 58, 60*
Money, T.   *60*
Montagna, W.   *180*
Montague, P.M.   *177*
Montgomery, H.   *176*
Moore, G.E.   *184*
Morais, A.   *119*
Moretti, A.   *117*
Morin, N.   *125*
Morita, H.   *58*
Morita, Y.   *121*
Morooka, N.J.   *120, 121*
Morrison, L.A.   *124*
Morrison, R.I.   *121*
Mortimer, P.H.   *122*
Morton, R.A.   *172*
Mosca, L.   *182*
Mosher, D.B.   *180*
Mott, N.F.   142, *180*
Mottaz, J.H.   *177, 180*
Moubasher, M.H.   *114*
Muhitch, M.J.   *129*
Mukherjee, M.   *173*
Mule, G.   *117, 126*
Muller, B.   *126*
Murata, M.   *59, 61*
Murgo, A.J.   *178*
Murphy, B.P.   *174*
Murthy, P.P.N.   *126*
Musajo, L.   167, *180*

Nadeau, Y.   *124–126, 128*
Nagai, K.   *123*
Nagata, S.   *180*
Nagayama, S.   *121*
Nakai, A.   58, *59*
Nakai, Y.   *123*
Nakamura, K.   *114*
Nakayasu, M.   *180*

Namikoshi, M.   *122, 123*
Nangia, A.   *124*
Napolitano, A.   *175, 180, 181*
Nathanson, L.   *180*
Naughton, G.K.   *180*
Nei, R.   *58*
Neish, G.A.   *114, 121*
Nelson, P.E.   73, 76–78, 96, *114, 125*
Nelson, R.M.   *180*
Nestler, H.J.   *57*
Nguyen, C.-D.   *126*
Nguyen, T.H.L.   *115*
Nicholson, J.M.   *177*
Nicholson, P.   *127*
Nicolaus, R.A.   140, 141, *180, 181*
Niessen, M.L.   *127*
Niimura, Y.   *59, 61*
Nikolakakis, A.   *124, 125, 128*
Nikulin, M.   *114, 118*
Nirenberg, H.   *125*
Nishino, H.   *116*
Nordlund, J.J.   *180*
Norris, G.L.F.   *113*
Nose, N.   *117*
Nozoe, S.   *114, 124, 125, 128*
Nummi, M.   *118*

Odell, W.D.   *182*
O'Donnell, K.   *115*
Ogasawara, M.   *60*
Ogren, L.   *182*
Ogunkoya, L.   *178*
Ohlrogge, J.B.   *127*
Ohokubo, K.   *121*
Ohshima, T.   *59*
Ohtsubo, K.   *118*
Oh-Uchi, T.   *57*
Oikawa, A.   *180*
Oikawa, H.   *116*
Okada, G.   *123*
Okamoto, Y.   *121*
Okuchi, M.   *116*
Okumura, H.   *114*
Okun, M.R.   153, *180*
Oliver, J.S.   *127*
O'Neill, A.B.   *62*
O'Neill, P.   *171*
Onoda, N.   *58*
Oriswold, J.R.   *185*
Ortonee, J.P.   *180*

Ortonne, J.P. *180*
Oshima, T. *61*
Oskarsson, A. *178*
Osterhout, M.H. *60*
Ostrovskii, M.A. *178*
Otsuka, T. *121*

Padwa, A. *59*, *60*
Pake, G.E. *174*
Palumbo, A. *175*, *177*, *180*
Palyusik, M. *115*, *122*
Panek, J.S. *59*
Panichanun, S. *57*
Panozishvili, K. *126*
Pare, J.R.J. *124*
Pargellis, C. *126*
Parikka, P. *114*
Park, J.J. *129*
Parker, G.H. *180*
Parrish, T.A. *181*
Parry, D.W. *127*
Parsons, P.G. *179*
Patakova-Juzlova, P. *125*
Pathak, M.A. *167*, *172*, *181*
Pathre, S.V. *115*, *116*, *122*
Patil, D.G. *174*, *181*
Pavanasasivam, G. *113*, *119*, *120*, *123*
Pawelek, J. *159*, *164*, *173*, *181*, *183*
Pawlosky, R.J. *115*, *121*
Payne, L.G. *61*
Pelli, B. *171*
Pena, N.B. *123*
Peranio, C. *181*
Perkowski, J. *120*
Perrone, G. *118*
Peterson, E. *179*
Peterson, E.W. *185*
Peterson, R.E. *120*
Pettijohn, D.E. *182*
Pfenninger, O.W. *182*
Philip, V. *120*
Piattelli, M. *181*
Piattoni-Kaplan, M. *62*
Pick, E.P. *61*
Pickford, G.E. *178*
Piekarski, S. *177*
Pietra, F. *119*
Pineiro, M.S. *114*
Pinkus, H. *183*
Pitot, H.C. *181*

Pitt, J.I. *115*
Pittner, R.A. *181*
Platt, H.W. *120*
Plattner, R.D. *115–118*, *120*, *124*, *125*, *127*, *129*
Plugge, P.C. *56*
Polis, B.D. *174*
Porta, G. *180*
Post, P.W. *174*
Potter, C.J. *58*
Powelek, J.M. *172*, *173*, *178*
Prelog, V. *2*, *56*
Price, J.E. *181*
Proctor, P. *179*, *181*
Proctor, R.H. *127–129*
Prota, G. *160*, *175*, *177*, *180*, *181*, *183*
Pryor, W.A. *175*
Pullman, A. *141*, *181*
Pullman, B. *141*, *181*
Punya, J. *123*
Pusset, J. *57*

Qiu, M. *58*
Quessy, S.N. *58*
Quevedo, W. *176*
Quevedo, W.C. *157*, *177*, *181*
Quevedo Jr, W.C. *177*, *181*
Quiclet-Sire, B. *59*

Rabie, C.J. *116*, *119*
Rakshit, R. *174*
Ramaiah, A. *181*
Ramakrishna, Y. *120*
Ramirez, E. *61*
Ramirez Luna, J.E. *62*
Randazzo, G. *116*
Rao, K.V. *174*
Rao, M.M. *120*, *123*
Raper, H.S. *139*, *157*, *181*, *182*
Rashatasakhon, P. *59*
Ravindranath, V. *120*
Raza, S.K. *121*
Razafimbelo, J. *57*
Redding, T.W. *182*
Reger, T.S. *60*
Reimann, E. *57*, *61*
Reisen, A. *124*
Remers, W.A. *182*
Repine, J.E. *182*
Reszka, K. *183*

Author Index

Rey, P. *56*
Rezanka, T. *125*
Rezanoor, H.N. *127*
Richardson, K.E. *116*
Rigby, J.H. *27, 59*
Riggs, N.V. *119*
Riley, F.C. *181*
Riley, P.A. *178, 182*
Rinehart Jr., K.L. *62*
Ripperger, H. *117*
Ritieni, A. *118*
Rivero, M.D. *61*
Rizzo, I. *122*
Robb, D.A. *182*
Roberts, R.G. *115*
Robertson, A. *172*
Robson, N.C. *182*
Rodighiero, G. *180*
Rödl, W. *58*
Rodriguez-Lopez, J.N. *182*
Roesslein, L. *124*
Rohwedder, W.K. *119, 120*
Rolinson, G.N. *117*
Romeo, N. *175*
Romer, A. *117*
Roquebert, M.-F. *119*
Rorsman, H. *182*
Rose, M. *59*
Rosel, M.A. *136, 182*
Rosengren, E. *182*
Ross, G.T. *182*
Rosso, M.L. *130*
Rottinghaus, A.A. *119, 125*
Rottinghaus, G.E. *114–116, 119, 121, 124, 125*
Roy, S. *161, 174, 182*
Roychowdhury, S.K. *174, 182*
Roychowdhury Jr, A. *166, 182*
Rozynov, B.V. *118, 120*
Rudowska, I. *167, 182*
Ruesch, M.E. *116*
Rullkotter, J. *117*

Sakai, K. *117*
Sakai, Y. *58, 59*
Sakamura, S. *116*
Sakuma, T. *120*
Salemme, J. *119*
Samanen, J.M. *61*
Samples, D. *123*

Samsonia, Sh. *57*
Sandberg, A.A. *184*
Sangai, M. *58*
Sano, T. *56, 58, 59, 61*
Sanson, D.R. *115*
Sarna, T. *176, 178, 182, 183*
Sasaki, M. *183*
Sato, K. *60, 61*
Sato, N. *117*
Sato, T. *59, 60*
Saul, R.L. *175*
Saunier, J.B. *59*
Sauriol, F. *115, 124–126, 128*
Sauvage, G.L. *56*
Savard, M.E. *114, 117, 124*
Sawano, M. *118*
Sayer, S.T. *114*
Schally, A.V. *182*
Schenkel, E.P. *130*
Schlemper, E.O. *115*
Schlesinger, W. *174*
Schmidt, R. *124*
Schoffelmeer, A.N.M. *62*
Schroeder, D.J. *118*
Schultz, T.M. *174*
Scott, J.A. *123*
Scott, P.M. *114, 117, 120*
Sealy, R.C. *174, 176–178, 182, 183*
Searles, S. *115*
Sehgal, V.N. *166, 183*
Seidel, P.R. *57*
Seifert, K.A. *114, 117*
Seiji, M. *167, 183, 184*
Semenst, V. *183*
Sen, M. *173*
Sever, R.J. *174*
Shamma, M. *57*
Shankland, D.L. *122*
Sharma, G.M. *60*
Sharma, S.K. *57*
Shaw, K.J. *115*
Shea, C. *181*
Shibata, K. *59*
Shibata, M. *60, 61*
Shibata, T. *183*
Shim, J.H. *127*
Shimizu, H. *57*
Shingu, Y. *129*
Shirota, T. *121*
Shizuri, Y. *124*

Shotwell, O.L. *120*
Shuttleworth, A. *116*
Sidebottom, P.J. *116*
Siems, K. *116*
Sigg, H.P. *114, 117, 121, 122*
Silvers, W.K. *172*
Simpson, R.T. *183*
Sinclair, J.B. *116*
Sinelair, R.S. *174*
Sinesi, S.J. *181*
Sinha Roy, S.P. *183*
Sioumis, A.A. *56*
Siverns, M. *126*
Slominski, A. *169, 183*
Smalley, E.B. *117, 120*
Smith, D.M. *118*
Smith, R. *61*
Smitka, T.A. *120–122*
Sneden, A.T. *124*
Snell, R.S. *172, 183*
Soares da Silva, N. *120*
Sobolev, V.S. *115, 118, 120*
Soboleva, N.A. *115, 124*
Solano, F. *172*
Solfrizzo, M. *116*
Soliman, S. *118*
Sorenson, W.G. *114, 118*
Soto-Hernandez, M. *61, 62*
Sparace, S. *125*
Spies, H.S.C. *121*
Spraul, M. *124*
Srivastava, A.K. *121*
Srivastava, R.S. *117, 121*
Stack, M.E. *123*
Stahelin, H. *114, 116*
Stahly, G.P. *113, 119, 123*
Staricco, R.J. *183*
Staudinger, H.J. *184*
Stea, G. *126*
Stegemann, J. *124*
Steglich, W. *57*
Steyn, P.S. *116, 119*
Stoll, Ch. *114, 121*
Stoltz, D.R. *117*
Stover, B.J. *173*
Stowell, C.P. *177*
St-Pierre, P. *125*
Streelman, D.R. *124*
Strong, F.M. *117–119*
Stuart, B.P. *115*

Sturua, M. *57*
Styriak, I. *120*
Subbarao, K.V. *174*
Sugawara, T. *123*
Sugiura, Y. *114, 121, 126*
Sugumaran, M. *183*
Suksamrarn, A. *61*
Sullivan, J.P. *62*
Sundheim, L. *115, 121*
Suzuki, K. *123*
Suzuki, T. *117*
Svoronos, P. *175*
Swan, G.A. *134, 136, 141, 159, 182–184*
Swanson, S. *62, 119, 126*
Swartz, H. *182*
Swenton, J.S. *59*
Sydenham, E.W. *118*
Szabo, G. *176, 177, 181, 184*
Szathmary, C.I. *115*
Szecsi, A. *126*

Taga, J. *58*
Tagami, H. *184*
Tait, D. *174*
Takahashi, M. *60*
Takahashi, T. *123*
Takatsuki, A. *128*
Takeda, K. *59*
Takeda, S. *59*
Takenouchi, K. *184*
Talmage, D.W. *182*
Tamm, Ch. *113, 114, 121, 122, 124, 126*
Tamura, O. *60*
Tamura, Y. *60, 61*
Tanaka, A. *123*
Tanaka, H. *57, 60, 61*
Tanaka, T. *57, 114, 117, 118, 121*
Tanticharoen, M. *123*
Tarin, D. *181*
Tatawawadi, S.V. *177*
Tateishi, Y. *57*
Tatsuno, T. *120, 121*
Taylor, A. *115, 117, 121, 124*
Taylor, S.L. *124, 125, 129*
Teetz, V. *58*
Tempesta, M.S. *114–116, 119, 121, 124, 125*
Tenconi, L.T. *184*
Terada, Y. *57*
Teshima, H. *116*

Thebtaranonth, Y. *123*
Theil, P.G. *118*, *121*
Thomas, M. *184*
Thomas, P.A. *121*
Thompson, A. *178*
Thomson, R.H. *184*
Thorn, G.W. *176*
Thrane, U. *125*
Tietze, L.F. *60*
Tjalve, H. *178*
Tjarks, L.W. *115*
Toda, J. *59*, *61*
Toda, K. *184*
Tokarnia, C.H. *124*
Tokuda, H. *116*
Tokumaru, Y. *117*
Tomioka, K. *116*
Tomita, Y. *183*, *184*
Tompsett, D. *184*
Toney, G.E. *116*
Tori, K. *123*
Toshima, S. *184*
Toussoun, T.A. *114*
Townsend, D. *178*
Townsend, J. *174*
Toyao, A. *60*
Tracy, J.K. *114*
Traldi, P. *171*
Trapp, S.C. *130*
Trapp, S.E. *129*
Truscott, G. *178*
Truscott, T.G. *174*
Tsakadze, D. *57*
Tsiakas, K. *175*
Tsuda, Y. *56*, *58*, *59*, *61*
Tsunoda, H. *117*, *121*
Tucci, S. *62*
Tudela, J. *182*
Tulloch, M. *126*
Turner, D.W. *58*
Turner, W.B. *116*
Tutel'yan, B.V. *120*
Tutel'yan, V.A. *115*, *118*
Tuthill, D. *123*

Udell, M.N. *128*
Udenfriend, S. *184*
Ueda, M. *173*
Ueno, I. *121*
Ueno, O. *121*

Ueno, Y. *114*, *116–118*, *121*, *126*
Ujszaszi, K. *123*
Ullrich, V. *184*
Umeda, M. *121*
Unna, K. *56*, *61*
Uotani, N. *123*

Van den Wijngaard, R. *178*
Vanderschuren, L.J.M.J. *62*
Van der Stap, J.G.M.M. *116*
Van de Ven, H.W.M. *62*
Vangedal, S. *115*
VanMiddlesworth, F. *124*, *127*, *129*
Van Oordt, G.J. *167*, *173*
Van Wyk, P.S. *118*
Varashin, M.S. *120*
Varkey, C. *172*
Varon, R. *182*
Varsavky, E. *122*
Vazquez, M.M. *62*
Venneste, W.H. *184*
Vepkhvadze, T. *57*
Vesonder, R.F. *117*, *119*, *120*
Viden, I. *125*
Vilhuber, H.G. *57*
Visconti, A. *116*, *117*, *119*
Vittimberga, B.M. *122*
Vittimberga, J.S. *122*
Vleggaar, R. *116*, *119*
Vogel, R.F. *127*
Voigt-Scheuermann, I. *125*
Voorhees, J.J. *184*
Vorovkov, A.V. *126*
Vrudhula, V.M. *118–120*

Waggott, A. *136*, *141*, *183*
Walaas, E. *184*
Waldmeier, F. *126*
Walser, M.M. *118*
Wang, S. *114*, *118–120*, *123*
Wang, Y. *126*
Wang, Z. *115*
Wanjala, C.W. *57*
Wanner, E.J. *183*
Ward, D.N. *173*
Ward, P.J. *118*
Wardeh, G. *62*
Warnes, H. *172*
Wasserman, H.H. *60*
Watanabe, A. *117*

Watanabe, H. *58*
Waterson, A.G. *60*
Weinstock, J. *56*
Weisleder, D. *116, 120, 129*
Wells, K.M. *123*
Wenzinger, G.R. *57*
Westerhof, W. *178*
Westling, M. *61*
Weston, G. *127*
Wheeler, M.H. *184*
White, E.P. *122*
White, R. *184*
Whitney, W.D. *184*
Whittern, D.N. *119*
Whyte, A.C. *123*
Widdowson, D.A. *58*
Wiesinger, D. *114*
Wilen, S.H. *61*
Wilkins, K. *125*
Williams, T.M. *123*
Willium, D.C. *172*
Wilson, M.C. *122*
Windels, C. *125*
Wink, J. *124*
Wolfe, L.A. *185*
Wolfram, L. *175*
Wolfram, L.J. *184*
Woo, L. *176*
Woode, M.K. *124*
Woods, M. *185*
Workman, R.J. *176*
Wright, M.R. *179*
Wrigley, S.K. *116*
Wu, Z. *127*
Wuchiyama, J. *129*
Wunderlich, J.A. *56*
Wurtman, R.J. *185*
Wyler, H. *136, 172*
Wyllie, L.M.A. *173*

Xu, Y.-C. *121*
Xue, Q. *127*

Yada, Y. *177*
Yagen, B. *119*
Yaijma, H. *182*
Yamaguchi, I. *128, 129*
Yamaoka, H. *178*
Yang, G. *127*
Yang, J. *115*
Yasunobu, K.T. *136, 185*
Yatawara, C.S. *114, 119, 122*
Yates, S.G. *119*
Ye, Y. *115*
Yoneyama, K. *128, 129*
Yoshimoto, T. *122*
Yoshinaga, M. *58*
Yoshizawa, T. *120, 121*
Young, C. *62*
Young, C.J. *114*
Young, T.E. *185*

Zajkowski, P. *124*
Zamir, L.O. *115, 124–126, 128*
Zander, J. *60*
Zard, S.Z. *59, 60*
Zarearo, R.M. *177*
Zehnder, M. *124*
Zeise, L. *185*
Zeitkiewicz, D.L. *117*
Zelickson, A.S. *177, 180*
Zenk, M.H. *20, 58*
Zhang, J. *115*
Zhang, L. *122*
Zhang, N. *115*
Zhang, Y. *59*
Zhou, Y. *114, 118*
Zhu, T.-X. *121*
Ziaiev, R. *57*
Ziegler, F.E. *124*
Zolnikova, N.Y. *126*
Zook, M. *125*
Zuberbuhler, A. *184*
Zurcher, W. *122, 124*
Zwick, W. *59*

Subject Index

Acetamide derivatives 43
Acetanhydride 34
Acetate 2, 99
Acetic acid 41
Acetone 38, 39, 45, 51, 52
Acetonitrile 25
3-Acetoxyscirpendiol 67
4-Acetoxyscirpendiol 67
15-Acetoxyscirpendiol 67
15-Acetoxyscirpendiol-4-β-glucoside 71
11-Acetylerysotrine 7
AcetylFS 4 93, 94
N-Acetyl-5-methoxytryptamine 155
3-Acetylneosolaniol 107
9-Acetylneosolaniol 70
Acetylroridin E 81, 86
Acetylroridin K 81, 87
8-AcetylT-2 tetraol 68
15-AcetylT-2 tetraol 67
AcetylT-2 toxin 72
3-AcetylT-2 toxin 108
Acetyltrichothecolone 75
4-Acetylverrol 71
Acetylverrucarin L 79, 84
3-Acetylvomitoxin 75, 99, 101, 102, 104–108
7-Acetylvomitoxin 75
15-Acetylvomitoxin 75, 106
Acid decarboxylation 136
Acremonium neo-caledoniae 71, 84, 86
Acremonium sp. 76, 96, 97
Acuminatin 69
1-Acyldihydroisoquinoline 44
Acyliminium 43
Addison's disease 155
Adrenaline 145, 155, 156
Adrenocorticotropin 155, 156
Alkaloids 2, 4, 18, 19
Alkenoids 5
Alkenyltrichloroacetamides 26
Amino acids 136
Aminohydroxyphenylalanine 137

Ammi majus 166
Ancymidol 101
Anthranilic acid 163
Antiasthmatic activity 54
Antibiotic activity 168
Antibiotic X379 81, 86
Antibiotic Y379 79, 84
Antifungal activity 168
Antileukemic activity 54
Antimelanocyte-antibodies formation 165
Antineoplastic activity 54
Aporphines 19
Apotrichodiol 92, 95, 113
3-*epi*-Apotrichodiol 92, 95
Apotrichodiols 113
Apotrichool 92, 95, 113
Apotrichothecenes 93, 103, 112, 113
Aromatic alkaloids 4
Aromatic hydrocarbons 138
Arylhydroindoles 31
N-Arylpropyl-enamides 46
Ascorbate 161, 162
Ascorbic acid 154, 161, 162, 169
Autoimmunization 165
16-Azaerythrinane derivatives 5
17-Azahomoerythrinane derivatives 5
Azodiisobutyronitrile 30, 34

Baccharin 82, 88
Baccharinoid B1 83, 89
Baccharinoid B2 83, 89
Baccharinoid B3 83, 89
Baccharinoid B4 82, 88
Baccharinoid B5 82, 88
Baccharinoid B6 82, 88
Baccharinoid B7 83, 89
Baccharinoid B8 82, 88
Baccharinoid B9 82, 88
Baccharinoid B10 82, 88
Baccharinoid B12 82, 88, 98
Baccharinoid B13 82, 88
Baccharinoid B14 82, 88

Subject Index

Baccharinoid B16   82, 88
Baccharinoid B17   82, 88
Baccharinoid B20   83, 89
Baccharinoid B21   82, 88
Baccharinoid B23   83, 89
Baccharinoid B24   83, 89
Baccharinoid B25   81, 88
Baccharinoid B27   82, 88
Baccharinoids   65, 88–90, 97, 98, 111, 112
Baccharinol   82, 88
Baccharis artemisioides   84, 86, 97
Baccharis coridifolia   65, 68, 72, 73, 84–86, 88, 89, 97, 112
Baccharis megapotamica   86, 88, 89, 97
Baccharis sp.   65, 97, 112
Baccharisol   83, 89
Bahah   164
Baras   164
Benzene   25, 27, 29, 32, 33, 37, 38, 40, 42
Benzophenanthridines   19
1,4-Benzothiazine   143
Benzothiazine isomers   145
Benzothiazole   143
Bioelectrets   142
Biological activity   2, 4, 21, 53, 63, 133, 155, 166
Birch reduction   40, 41
Bisbenzylisoquinolines   19
Bischler-Napieralski reaction   36
1,3-Bis(diphenylphosphino)propane   26
1,3-Bis(trimethylsilyloxy)butadiene   35
Bohak   164
Boronic acid   49
Bromoacetic acid ester   31
2-Bromohomoveratrylamine   34
2-Bromopiperonal   49
Bruylants reaction   28, 31
Bufo melanostictus   167
8-Butyrylneosolaniol   71

Caffeine   155
Calonectria nivalis   67, 68
Calonectrin   68, 106, 107, 109
Carbon   135
Carboxyls   148
Cardiovascular activity   55
Catechol   137, 154, 165
Catecholamine   165
Catecholamines   143–145, 171
Catecholase activity   152, 153

Catechols   138
Caucasoid skin   151
CBD_2 76
Cephalezomine M   15, 16
Cephalosporium crotocinigenum   68
Cephalosporium sp.   96
Cephalotaxus fortunei   15
Cephalotaxus harringtonia var. *nana*   15
Cephalotaxus sp.   54
Ceratopycnidium baccharidicola   84, 86, 97, 112
Ceratopycnidium sp.   97
Cercophora areolata   86
Cercophora sp.   97
Cesium fluoride   44
i-Bu-Chloroformate   37
Chloroformylacetate   36
Chlorpromazine   145, 149
N-Chlorsuccinimide   27
Cholecalciferol   155
Cinnamyltrichothecolone   76
Claisen condensation   34
Cloudman melanoma   159, 167
Cocculidine N-oxide   12
Cocculolidine   5
(±)-Cocculolidine   23
Cocculus laurifolius   12
Comosine   50, 51
Congenital vitiligo   164
Copper(II) acetate   43
Copper(II) triflate   43
Coral tree   3
Corticosteroids   168, 169
Cresolase activity   152, 153
Crotocin   68, 98
Curare alkaloids   53, 55
Curare-like activity   2
Curie point pyrolysis   137
Cuttlefish   150
Cyclic-AMP   156
Cyclic-GMP   156
Cycloalkanoylalkylamines   31
Cyclohexanoylglyoxylic acid   24
Cyclohexanoylacetic acid   25
Cyclohexanylacetaldehyde   29
10,13-Cyclo-trichothecanes   90
Cylindrocarpon sp.   85, 86, 97, 98
Cylindrocladium floridanum   67
Cylindrocladium sp.   96, 97
Cysteic acid   135

Cysteine 169
Cysteinyldopa 145, 154
5-S-Cysteinyldopa 137
Cysteinyldopa-quinones 145
Cytochrome 149
Cytochrome P450 102, 106

4-DeacetoxyT-2 toxin 70
3-Deacetylcalonectrin 67
15-Deacetylcalonectrin 67, 106, 107
4-DeacetylT-2 toxin 107
15-DeacetylT-2 toxin 70
Decaline 40
12,13-Deepoxy-roridin E 90
Dehydroapotrichodiol 113
3-Dehydroapotrichodiol 91, 95
2′-Dehydroverrucarin A 79, 84
Deionized water 134
(±)-3-Demethoxy-1,2-dihydrocomosidine 52, 53
(±)-3,17-Demethoxy-1,2-dihydrocomosidine 50
3-Demethoxy-1,2-dihydroerysotramidine 24
3-Demethoxy-1,2-dihydroerysotrine 24
3-Demethoxyerysotramidine 26
(±)-Demethoxyerythratidinone 33
3-Demethoxyerythratidinone 27, 33, 34, 37, 39, 40
3-Demethoxy-Δ^3-erythratidinone 40
(−)-3-Demethoxyerythratidinone 24, 25
(±)-3-Demethoxyerythratidinone 32, 44, 45
Demethoxytetrahydroerysotramidine 39
(−)-3-Demethoxytetrahydroerysotramidine 25, 26
18-De-O-methylholidine 14
Dendrodochium toxicum 96
Dendrostilbella sp. 67, 96
8-Deoxotrichothecin 68
8-Deoxotrichothecinol A 65, 68
12,13-Deoxydiacetoxyscirpenol 77
C16-Deoxygenated erythrinane alkaloids 18
12,13-Deoxy-(2″E)-isotrichoverrin B 77
4-Deoxynivalenol 75, 98
7-Deoxynivalenol 75
12,13-Deoxyroridin E 81, 85
Deoxysambucinol 113
3-Deoxysambucinol 90, 91, 95, 113
12,13-Deoxytrichodermadiene 77
12,13-Deoxytrichoverrin A 77

12,13-Deoxytrichoverrin B 77
12,13-Deoxytrichoverrins 112
7-Deoxyvomitoxin 75
D-*seco*-Derivatives 5
Dess-Martin periodane 26
Dexamethasone 170, 171
4β,15-Diacetoxy-10,13-cyclotrichothecan-9α,12-diol 73
Diacetoxyscirpenol 68, 96, 108, 109
3,4-Diacetoxyscirpenol 68
3,15-Diacetoxyscirpenol 68, 107
4,15-Diacetoxyscirpenol 68, 105, 107, 108
Diacetylneosolaniol 71
4,15-Diacetylnivalenol 76, 108
Diacetylsambucinol 90, 93, 95
Diacetylverrucarol 65, 68
3,15-Diacetylvomitoxin 76
Diastereoisomers 65, 73, 77, 78
Diazepam 54
Diazomethane 31
Dibenzazecine 49
Dibenzazecine alkaloids 21
Dibenzazonine 18, 19
Dibenzazoninedione 19
Dibutyryl cyclic-AMP 156
Dideacetylcalonectrin 106
Diels-Alder addition 30, 31, 51
Diels-Alder reaction 35, 45, 46
Dienoids 5
Diepoxyroridin H 80, 85
Dihydro-β-erythroidine 53, 54
2,7-Dihydrohomoerysotrine 49, 55
Dihydronoradrenaline 149
Dihydrophenylnoradrenaline 149
Dihydroschelhammeridine 50, 51
8-Dihydrotrichothecinol A 73
2α,13-Dihydroxyapotrichothecene 113
4,4′-Dihydroxybiphenyl 163, 170
7,8-Dihydroxycalonectrin 69, 106
7,8-Dihydroxy-15-deacetylcalonectrin 68
7,8-Dihydroxydiacetoxyscirpenol 69
5,6-Dihydroxyindole 133, 136–138, 140, 158–160, 163, 170
5,6-Dihydroxyindole-2-carboxylic acid 137, 158–160, 170
5,6-Dihydroxyindole-4,7-dicarboxylic acid 137
7,8-Dihydroxyisotrichodermin 67
3,12-Dihydroxy-13-methoxy derivative 49
5,6-Dihydroxy-2-methylindole 145

Dihydroxyphenylalanin 152
5,6-Dihydroxyphenylalanin 133, 140
15,16-Dimethoxy-2,8-dioxoerythrinane 42–44
15,16-Dimethoxyerythrinane 28
cis-15,16-Dimethoxyerythrinane 34
15,16-Dimethoxy-4-oxoerythrinane 46
15,16-Dimethoxy-4-oxo-$\Delta^{6,7}$-erythrinane 46
15,16-Dimethoxy-8-oxoerythrinane 42, 43
(S)-(+)-3,4-Dimethoxyphenylalanine-ester 37
3,4-Dimethoxyphenyl-L-alanine ester 36
(S)-(+)-Dimethoxyphenylalanine methylester 24
6,7-Dimethoxy-2-tetralone 53
Dimethyl-n-dodecyl-n-undecyl ammonium hydroxide 135
Dimethyl(methylthio) sulfonium tetrafluoroborate 43
Dimethyl phosphonoacetyl chloride 33
Dioxane 34, 48
(+)-10,11-Dioxoerysotrine 7
10,11-Dioxoerythraline 8
2,8-Dioxoerythrinane 36
7,8-Dioxoerythrinane 24, 27
2,8-Dioxohomoerythrinane 49
2,8-Dioxo-16-hydroxy-15-methoxyerythrinane 41
Dioxopentenoic acid ester 37
Dioxopyrroline dienophiles 51
Dioxopyrrolobenzazepine 50, 51
Dioxopyrroloisoquinolines 35, 50
Dioxoschelhammerane 50
2,8-Dioxoschelhammerane 51
Dioxygenase activity 168
Dioxygenase reaction 161
Dioxygenases 169
15,16-Dioxygenated-8-oxo-erythrinanes 42
o-Diphenol 153
Diterpenoids 73, 76, 93
Diuretic activity 54
Dopa 145, 151–155, 158–160, 165, 169–171
L-Dopa 138, 169
Dopachrome 137, 154, 158–160, 170
L-Dopachrome 159
Dopachrome conversion factor 159, 160
Dopachrome isomerase 159
Dopachrome oxidoreductase 159
Dopachrome tautomerase 159, 170

Dopachromes 145
Dopamelanin 136, 137, 158, 159
D,L-Dopa melanin 148
Dopamine 159
Dopamine-melanin 137
Dopa oxidase 169
Dopaquinone 152, 154, 155, 158, 160
Dopaquinoneimine conversion factor 159
Dopaquinones 145
Dysazecine 21
Dyshomoerythrine 54
Dysoxyline 21
Dysoxylum lenticellare 16, 17

Electrochemical oxidation 144
Electron spin resonance spectroscopy 139, 146, 148–150, 167
Electron-transfer agents 149
Electrooxidation 143
Enzyme activity 151, 154, 157, 167
12-Epi-apotrichothecene 113
Epidermal melanin synthesis 150
6′-Epi-13′-epiroridin A 81, 86, 111
11-Epi-12-epitrichothecenes 90
(+)-Epierythrinine 10
13′-Epiisororidin E 80, 86, 111
11-Epiisotrichodiol 113
13′-Epiroridin E 81, 86
9-Epitrichodiol 102
9-Epitrichotriol 102
7β,8β-Epoxide 73
7β,8β-Epoxyisororidin E 80, 85
9β,10β-Epoxyisotrichoverrin A 78
9β,10β-Epoxyisotrichoverrin B 78
1,6α-Epoxyrobustivine 14
7β,8β-Epoxyroridin H 80, 85
12,13-Epoxytrichodiene 102
12,13-Epoxytrichothecane 64
12,13-Epoxytrichothecene 66, 101, 102, 104, 105, 107, 113
12,13-Epoxytrichothec-9-enes 65, 66–73
12,13-Epoxytrichothec-9-en-8-ones 65, 75, 76, 97
Erymelanthine 5
Erysodienone 18, 19
($\pm$)-Erysodienone 40
Erysodienonemethylether 35
Erysodine 2, 5, 54
Erysopine 2, 5
Erysotramidine 36, 38

Subject Index

(+)-Erysotramidine 36, 37
(±)-Erysotramidine 26, 45, 47
Erysotrine 30, 31, 36, 38
(+)-Erysotrine 36, 37
Erysovine 2, 5
(+)-Erythbidin B 9
Erythema 167
Erythraline 2, 5, 19, 20, 36
Erythramine 2, 5
Erythratine 2, 5
Δ^3-Erythratinone 20
Erythrina alkaloids 4, 5, 18, 21, 53, 55
Erythrina americana 53, 54
Erythrina bidwillii 8, 9
Erythrina caffra 10
Erythrina crista-galli 2, 3
Erythrina latissima 7, 11
Erythrina poeppigiana 13
Erythrina sp. 2, 4, 53, 55
Erythrina stricta 7
Erythrina suberosa 54
Erythrina variegata 6, 54
Erythrina velutina 54
Erythrinane 28
($3R,5S,6S/3S,5R,6R$)-Erythrinane 4
Erythrinane alkaloids 2, 4, 6–13, 18–23
Erythrinane derivatives 25, 31, 43
Erythrinanes 4, 5, 18, 20, 22–24, 28, 30, 31, 33, 36–38, 41, 45, 46, 50, 55
(+)-Erythrinine 13
(±)-2-*epi*-Erythrinitol 27, 28
Erythroidine 2, 5, 53
α-Erythroidine 53, 54
β-Erythroidine 53, 54
Erythromotidienone 6
Erythrosotidienone 6
Escherichia coli 101
Estrogen 155, 156
Estrogenic activity 168
Ethanol 136
Ether 162
Ethyleneglycol 37, 51
3-Ethylpentane-3-thiol 38
Eumelanin 140, 157, 160
Eumelanins 134–136, 138, 139, 146, 159
Eye-melanin 134

F-11703-1 76
F-11703-2 76
Farnesyldiphosphate 99

2-*trans*-6-*trans*-Farnesyldiphosphate 99
Female hormones 155
Folic acid 154
Fortunine 15
Friedel-Crafts acylation 31
Friedel-Crafts method 50
FS 1 91, 95, 112
FS 2 92, 94
3-*epi*-FS 2 92, 94
FS 3 91, 95
FS 4 91, 94, 112
Fungal macrocycles 97
Furocoumarins 166
Fusarenone 75, 108
Fusarenone X 76
Fusarium acuminatum 66–72, 75, 76, 96
Fusarium acuminatum (*heterosporum*) 66, 67, 70, 71
Fusarium acuminatum (*sulphureum*) 69–71
Fusarium acuminatum subsp. *armeniacum* 96
Fusarium avenaceum 69, 75, 76
Fusarium avenaceum (*roseum*) 71
Fusarium camptoceras 66, 67, 69, 75, 76
Fusarium chlamydosporium 96
Fusarium chlamydosporium (*tricinctum*) 70
Fusarium compactum 96
Fusarium crookwellense 66–69, 72, 75, 76, 91, 92, 98, 109
Fusarium culmorum 66–69, 71, 72, 75, 76, 91, 92, 98, 99, 101, 102, 104, 106, 108, 113
Fusarium equiseti 66–72, 75, 76, 96, 108
Fusarium equiseti (*avenaceum*) 67, 68
Fusarium equiseti (*compactum*) 69, 70
Fusarium equiseti (*concolor*) 67, 68
Fusarium equiseti (*roseum*) 68
Fusarium equiseti (*semitectum*) 67, 68
Fusarium graminearum 66, 67, 69, 71, 72, 75–77, 91, 92, 97–99, 101, 104, 106, 108, 109
Fusarium graminearum (*decemcellulare*) 69
Fusarium graminearum (*roseum*) 67–70, 72, 75, 76
Fusarium graminearum (*tricinctum*) 68, 70, 72
Fusarium lateritium 69
Fusarium moniliforme 69, 71–73, 75, 76
Fusarium nivale 75, 96
Fusarium oxysporum 66, 69, 71, 72, 75
Fusarium oxysporum (*lateritium*) 69, 76
Fusarium poae 66, 67, 69–72, 75, 76, 91, 101

Fusarium poae (tricinctum) 69
Fusarium sambucinum 66–72, 75, 76, 91–93, 96, 98, 99, 101, 107–109, 112
Fusarium sambucinum (roseum) 68–70, 72
Fusarium sambucinum sensu lato 96
Fusarium sambucinum sensu stricto 96
Fusarium sambucinum (sulphureum) 67, 68, 70, 71, 75, 76
Fusarium semitectum 69, 71, 75, 76
Fusarium semitectum (roseum) 66, 67
Fusarium solani 69, 71, 72, 75
Fusarium sp. 70, 73, 76, 90, 93, 96–98, 101, 102, 106, 108
Fusarium sporotrichioides 66–72, 75, 76, 91–93, 97, 99, 101, 102, 104, 106–109, 111
Fusarium sporotrichioides (episphaeria) 75
Fusarium sporotrichioides (moniliforme) 71, 75
Fusarium sporotrichioides (nivale) 75, 76
Fusarium sporotrichioides (oxysporum) 75, 76
Fusarium sporotrichioides (poae) 69–72
Fusarium sporotrichioides (solani) 68–71
Fusarium sporotrichioides (tricinctum) 68–71, 75
Fusarium stilboides 72
Fusarium subglutinans 72, 75
Fusarium torulosum 96
Fusarium tricinctum 96
Fusarium tricinctum sensu stricto 96
Fusarium tumidum 69
Fusarium venenatum 66, 67, 69, 91, 92, 96

Gas chromatography 138
Gliocladium sp. 96, 97
Gliocladium virens 66, 67
(+)-15β-D-Glucoerysopine 11
(+)-16β-D-Glucoerysopine 11, 12
Glucosamine 136
Glutamate oxidase 152
γ-Glutamyl transpeptidase 154
Glutathione 154
Glutathione reductase activity 154
Glutinosin 64
Goldfish tyrosinase 169
Gramilaurone 90, 91, 95
Growth inhibition activity 168

Harding-Passey mouse melanoma 169
Harzianum A 71

Heck reaction 27
Heterocyclic compounds 150
8-n-Hexanoylneosolaniol 72
5-Hexenoyl chloride 44
1-Hexynoyldihydroisoquinoline 44
Holarrhena floribunda 65, 68, 75, 76, 91
Holidine 5
Homoerysodienone 47
C-Homoerysodienone 48
Homoerythrina alkaloids 4
($\pm$)-Homoerythrinadienone 48
B-Homoerythrinane 32
Homoerythrinane alkaloids 4, 14–17, 20, 21, 31
Homoerythrinanes 4, 5, 18, 45, 49, 54, 55
S-(+)-Homolaudanosine 21
Homoveratrylamine 24, 25, 27, 37
Homoveratrylimide 45
Homoveratryl isonitrile 44
Hormones 155
Hydriodic acid 137
Hydrocortisone 156
Hydrogen 136
Hydroquinone 154, 170, 171
3-Hydroxyanthranilic acid 163, 168
2α-Hydroxyapotrichothecenes 113
7-Hydroxycalonectrin 69
8-Hydroxycalonectrin 69, 106
7-Hydroxydiacetoxyscirpenol 69
15-Hydroxy-12,13-epoxytrichothecene 104
5-Hydroxyindole acetic acid 168
8α-Hydroxyisororidin E 81, 86
7-Hydroxyisotrichodermin 67
8-Hydroxyisotrichodermin 67
15-Hydroxyisotrichodermin 106
7-Hydroxyisotrichodermol 66
8-Hydroxyisotrichodermol 66
8α-Hydroxyisotrichodiol 92, 94
8α-Hydroxyisotrichotriol 92, 94
8β-Hydroxyisotrichotriol 92, 94
16-Hydroxyisotrichotriol 92, 94
8-Hydroxyisotrichoverrin A 72
8α-Hydroxyisotrichoverrin A 112
2α-Hydroxylenticellarine 16
16-Hydroxy-15-methoxy-3-oxoerythrinane 41
p-Hydroxyphenylpyruvic acid 163, 170, 171
p-Hydroxypropiophenone 154
N-Hydroxypyridinethione-Na 37

3-Hydroxypyrrol-4,5-dicarboxylic acid   163
8β-Hydroxyroridin E   87
12′-Hydroxyroridin E   111
16-Hydroxyroridin L-2   72
6β-Hydroxy-rosenonolactone   73, 76, 93
8α-Hydroxysambucoin   91, 95
8β-Hydroxysambucoin   91, 95
3-epi-12-Hydroxyschelhammericine   55
8-Hydroxyscirpene   66
8β-Hydroxyscirpene   97
3′-HydroxyHT-2 toxin   71, 73
3′-HydroxyT-2 toxin   72, 73
16-Hydroxytrichodermadienediol A   71
16-Hydroxytrichodermadienediol B   71
7-Hydroxytrichodermol   66
2α-Hydroxytrichodiene   91, 94, 102
11α-Hydroxytrichodiene   91, 94, 102
15-Hydroxytrichodiene   93
16-Hydroxytrichodiene   92, 94
8-Hydroxytrichothecenes   104
16-Hydroxytrichothecenes   104
3′-HydroxyT-2 triol   70
5-Hydroxy tryptamine   168
12′-Hydroxyverrucarin J   84
Hyperpigmentation   155
Hypnotic activity   54

Imidazole   38, 52
Indican   168
Indole   163, 170
Indoleamine-2,3-dioxygenase   168, 169, 170, 171
Indole-5,6-quinone   140, 141, 158, 160
Indoles   138
Indoline-carboxylic acid   141
Indolylpyruvic acid   163, 170
Insecticidal activity   63
Intermedine   155
Iodophenethylamines   29
Isatin   163
IsoBaccharin   82, 88
IsoBaccharinol   82, 88
IsoBaccharisol   83, 89
8-Isobutyrylneosolaniol   71
Isocrotonyltrichodermol   68
Isomeric alkaloids   53
IsoMiotoxin D   82, 88
Isoprenaline   145
Isororidin A   81, 86
Isororidin E   80, 86, 111

Isororidin K   81, 86
Isosatratoxin F   80, 85
Isosatratoxin G   80, 85
M Isosatratoxin H   80, 85
S Isosatratoxin H   80, 85
IsoT-2 toxin   72
Isotrichodermin   67, 96, 104, 106, 109
Isotrichodermol   66, 101, 102, 104–106, 109
Isotrichodiol   92, 94, 101, 102, 104, 113
Isotrichool   92, 94, 102, 113
Isotrichothecin   90
Isotrichotriol   92, 94, 102
Isotrichotriols   104
Isotrichoverrin A   72
(2″E)-Isotrichoverrin A   72
(2″E,4″Z)-Isotrichoverrin A   72
Isotrichoverrin B   72
(2″E)-Isotrichoverrin B   72
(2″E,4″Z)-Isotrichoverrin B   72
Isotrichoverrin C   73
Isotrichoverrins   111
Isotrichoverrol A   71
(2′E)-Isotrichoverrol A   71
Isotrichoverrol B   71
(2′E)-Isotrichoverrol B   71
15-IsovalerylT-2 tetraol   70
Isoverrucarol   66, 106

Keratinocytes   151, 155, 165, 166
Kilas   164
Kynurenine   168

Lactonic alkaloids   5
Lauroyl peroxide   43, 44
Lenticellarine   54
L-Leucine   108
Leucodopachrome   158, 160
Lewis acid   24
Liquid chromatography   143
Loukacinol A   91, 95
Loukacinol B   91, 95
Loukacinols   97

Macrocycles   64, 65, 90
Malignancy   149
Malpighian cells   157, 167
Manganese(III) acetate   43
Mass spectrometry   138
Meladenin   166

Melanin   140, 158, 160, 165, 170
Melanin of sepia   150
Melanin pigmentation   146
Melanins   132–139, 142, 145–150, 154, 157
Melanin synthesis   154
Melanin-stimulating hormones   154
Melanochrome   158–160
Melanocytes   133, 150, 151, 155, 157, 159, 165–167
Melanocytotoxicity   164, 165
Melanogenesis   136, 139, 143, 152–154, 156, 158–162, 165, 167, 168
Melanoma cells   149, 150, 153, 169
Melanoproteins   135, 136
Melanosoma   152
Melanosomal tyrosinase   169
Melanosomes   134, 135, 142, 151, 152, 157, 165, 167
Melanotropin   159
Melatonin   149, 155, 156
Memnoniella echinata   66, 67
Memnoniella sp.   96, 97
Mercaptamine derivatives   154
Methanol   145
15-Methoxyerythrinane   31, 32
2α-Methoxylenticellarine   17
Methoxyschelhammericines   51
1-Methoxy-3-trimethylsilyloxybutadiene   35
Methyl acrylate   50
4-Methylcatechol   137
α-Methylcyclodopa   145
α-Methyldopa   144
α-Methyldopachrome   145
α-Methyldopa-quinone   144
Methylene chloride   25
12,13-Methylenedioxydibenzazecine   48, 49
15,16-Methylenedioxyerythrinane   28
15,16-Methylenedioxy-8-oxoerythrinane   30, 39, 42
cis-15,16-Methylenedioxy-8-oxoerythrinane   29
trans-15,16-Methylenedioxy-8-oxo-erythrinane   29
15-*O*-Methylerysodienone   35
Methyl (*E*)-7-iodo-2-heptenoate   53
N-Methylmorpholine   37
O-Methyltaxodine   21
Mevalonate   99
[3,4-$^{13}C_2$]-Mevalonate   99
Mevalonic acid   98

Mevalonic acid lactone   100
[2-^{13}C]-Mevalonic acid lactone   99
Michael addition   29, 37
Michael type addition   18
Microdochium nivale   75
Microdochium sp.   96
Miophytocen A   82, 88
Miophytocen B   82, 88
Miophytocens   90
Miotoxin A   81, 88
Miotoxin A 13′-glucoside   81, 89
Miotoxin B   82, 88
Miotoxin C   82, 89
Miotoxin D   73, 82, 88
Miotoxin E   82, 88
Miotoxin F   82, 89
Miotoxin F glucoside   82, 89
Miotoxin G   65, 72
Mitochondria   152
Mixed function oxidases   152, 161
Molluscicidal activity   54
Monascus purpureus   91, 97
Monoxygenase reaction   161
Monoxygenases   161
8-MOP   166, 167
Morphinanes   19
α-MSH   155, 156
β-MSH   155, 156
Muconomycin A   79, 84
Muconomycin B   79, 84
Mushroom tyrosinase   152
Myrothecium leucotrichum   84
Myrothecium roridum   66, 71–73, 84–86, 109
Myrothecium sp.   86, 96–98, 101, 102, 111, 112
Myrothecium verrucaria   68, 70–72, 77, 78, 84–87, 96, 111, 112
Myrotoxin A   79, 84
Myrotoxin B   79, 84
Myrotoxin C   79, 84
Myrotoxin D   79, 84
Myrotoxins   84, 90, 112
Mytoxin A   81, 85
Mytoxin B   80, 85
Mytoxin C   81, 85
Mytoxins   90

Naphthalenoisoquinoline   47
Natural skin tanning   145
Neoproaporphine   19, 47

Neoproaporphine derivative 18
Neosolaniol 69, 73, 96, 107–109
Neosporol 93
Nerolidyldiphosphate 99
Nerve cells 165
Neuromelanin 142
Neuromuscular blocking activity 2
Nicotiana tabaccum 92, 93, 101
Nitrogen 20, 135
5-(Nitromethyl)-1,3-benzodioxole 29
Nitroxide radicals 148
Nivalenol 75, 98, 108
Non-aromatic alkaloids 4
Noradrenaline 145, 155
Noramurine 19, 20
A-Norerythrinane 32
Norisosalutaridine 19, 20
(*S*)-Norprotosinomenine 18, 19
(*S*)-Norreticuline 19, 20
(*S*)-[1-^{13}C]-Norreticuline 20
A-Norschelhammerane 32
NT-1 69
NT-2 68
Nucleic acid 157

Olefins 161
Oxalylchloride 24
Oxene mechanism 161
Oxinoid species 161
8-Oxocalonectrin 76
2-Oxocyclohexane carboxylic derivatives 24
8-Oxo-15-deacetylcalonectrin 75
8-Oxodiacetoxyscirpenol 76
8-Oxoerythralidine 36
8-Oxoerythraline-epoxide 8
2-Oxoerythrinane 34
$\Delta^{6,7}$-Oxoerythrinane 44
8-Oxo-$\Delta^{6,7}$-erythrinane 27
Oxoerythrinanes 42
4-Oxoerythrinanes 44, 46
8-Oxoerythrinanes 39
10-Oxoerythrinanes 39
cis-11-Oxoerythrinanes 31
(+)-8-Oxo-α-erythroidine epoxide 13
8-Oxoisotrichodermin 75
Oxygen 139, 145, 146, 161, 162

Palmitylscirpentriol 73
PalmitylT-2 tetraol 73
Palmityltrichothecolone 76

Pantothenic acid 154
Papain 134
Paper chromatography 137
Pavines 19
PD 113325 79, 84
PD 113326 80, 85
8-Pentenoylneosolaniol 71
8-*n*-Pentanoylneosolaniol 71
Peptides 136
Peracetic acid 135
Peroxidase 153
Phaeomelanins 133, 134, 136, 137, 139, 143, 146
Pharmacological activity 53, 55
Phelline comosa var. *robusta* 14
Phenethylamine 40
Phenethyl-cyclohexanylethylamines 28
N-Phenethylhydroindole derivatives 23, 24
1-Phenethylisoquinoline 47
1-Phenethylisoquinoline derivatives 4
Phenethylisoquinolines 21
S(+)-1-Phenethyltetrahydroisoquinoline 21
Phenolic groups 148
Phenols 138
Phenylacetic acid 40
Phenylalanin 133
Phenylhydroindoles 31
Phenylpyrroles 30
Phenylselenoacetaldehyde 34
Phenyl vinylsulfoxide 29
Phoma sp. 84, 97
Phomopsis leptostromiformis 86
Phomopsis sp. 97
Phospholipid-P 152
Phosphoric acid 24
Photobleaching 147
Phytotoxic activity 63
Pictet-Spengler reaction 23, 32
Pigmented neurons 142
Pigment epithelium–choroid complex 147
Pigment forming activity 151
Pituitarin 156
Poikilopolymer 141
Polyphosphoric acid 39
Potassium cyanide 28
Potassium permanganate 137
Premelanosomes 151
Progesterone 156
Prolactin 155
Pronase 134

2-Propanol 27
8-Propionylneosolaniol 70
4-PropionylHT-2 toxin 72
i-Propylmagnesium chloride 29
Prostaglandin 156
Protoberberines 19
Protyrosinase 151
Psoralea corylifolia 166
Psoralen 166–169
Pulse radiolysis 145
Pummerer reaction 29, 31, 42, 45
PUVA therapy 166
Pyridine derivatives 154
Pyrimidine 167
Pyrrole-2,3-dicarboxylic acid 137
Pyrroles 138
2,3,4,5-Pyrroletetracarboxylic acid 137
Pyrrole-2,3,5-tricarboxylic acid 137
Pyrrolic acids 137
Pyrrolobenzazepines 37
Pyrroloisoquinoline carboxylic acid ester 37
Pyrroloisoquinolines 31, 35

Raper-Mason scheme 143, 158, 159, 170, 171
Rayleigh scattering of Mössbauer radiation 138
(*S*)-Reticuline 19
Roridin A 65, 81, 86, 87, 96, 97, 111
Roridin A glucoside 81, 89
Roridin C 66
Roridin D 65, 81, 86, 97
Roridin D glucoside 81, 89
Roridin E 65, 81, 84, 86, 87, 89, 97, 110
Roridin E glucoside 81, 89
Roridin E-2 81, 86
Roridin H 65, 80, 85
Roridin J 80, 85
Roridin L-2 72, 111
(2′*E*)-Roridin L-2 72
Roridin relatives 85–87
Roridins 65, 85–87, 90, 98, 111, 112
Roritoxin A 80, 85
Roritoxin B 80, 85
Roritoxin C 79, 85, 90, 98
Roritoxin D 79, 85
Roritoxins 90
Rosenonolactone 73, 76, 93
Rosololactone 73, 76, 93
Russula nigricans 133

Sambucinic acid 91, 94
Sambucinol 90, 91, 95, 112, 113
Sambucoin 91, 95
Satratoxin C 79, 84
Satratoxin D 81, 86
Satratoxin F 80, 85
Satratoxin G 80, 85
Satratoxin H 80, 85
Satratoxin H isomer 80, 85
Satratoxins 90, 112
(5*S*,6*S*/5*R*,6*R*)-Schelhammerane 4
Schelhammerane intermediates 37
Schelhammeranes 4, 22, 32, 46
Schelhammericine 5, 50, 52
epi-Schelhammericine 21, 52
3-*epi*-Schelhammericine 21, 54
Schelhammeridine 5, 50, 51
epi-Schelhammeridine 51
Schmidt rearrangement 53
Scirpane 64
Scirpene 66
Scirpene-3,4-diol 106
Scirpen-3,7,8,15-tetraol 67
Scirpentriol 66
Scirpen-3,4,8-triol 66
Scirpen-3,4,15-triol 66
Selaginoidine 5
Sepiomelanin 134–137
Sesquiterpenoid epoxides 63
Sex hormones 155, 156
Shweta Kustha 164
SIPI-299-B 79, 84
SIPI-299-O 79, 84
Skin cancer 147
Skin pigmentation 155–157
Sodium borohydride 137
Sodium hydroxide 137
Solaniol 73
Solulene 100 135
Spasmolytic activity 54
Spicellum roseum 66, 68, 96
Spicellum sp. 96
Spin trapping 146
Spiro-2-tetralones 52
Sporol 91, 95, 113
Sporotrichiol 70
Stachybotrys albipes 71, 84
Stachybotrys alternans 96
Stachybotrys atra 66, 70–72, 84–86, 91, 96, 97, 101

Subject Index

Stachybotrys chartarum 96
Stachybotrys cylindrospora 66, 67
Stachybotrys kampalensis 71, 84–86
Stachybotrys microspora 66, 71, 84–86
Stachybotrys sp. 96, 97, 101
Strecker reaction 28, 31
Substantia nigra 149
N-Substituted 1-acyldihydroisoquinolinium derivatives 44
C6-Substituted C5-spiroisoquinoline derivatives 35
N-Substituted C5-spiroisoquinoline derivatives 32
Succinoxidase 152
Sucrose density gradient ultracentrifugation 134
Sulfonic acid 27
Sulfur 135
Sulfur compounds 154
Sulfuric acid 41
Superoxide 146
Suzuki coupling procedure 49
Swern oxidation 32, 33
Synthetic D,L-melanin 150

Taurine 135
Testosterone 156
Tetrahydrofuran 25
Tetrahydroisoquinoline 143
T-2 Tetraol 66, 107
Thiamine 155
Thionium 43
Titanium tetrachloride 25
Tobacco 101
Toluene 26–31, 38, 39, 45, 49, 52, 135
p-Toluenesulfonic acid 43
Toxic activity 63
Toxicity 149
HT-2 Toxin 70, 107
T-2 Toxin 71, 96, 99, 102, 104, 105, 107–109
Triacetoxyscirpene 70, 107
Tri-n-butyllithium-thiostannane 34
Tri-n-butylstannane 30, 34
2,3,4-Tricarboxylic acid 137
Trichloroacetylchloride 27
Trichoderma harzianum 71
Trichoderma lignorum 96
Trichoderma polysporum 66, 67
Trichoderma sp. 96, 97, 101
Trichoderma viride 67, 96

Trichoderma viride (lignorum) 72
Trichodermadiene 71
Trichodermadienediol A 71
Trichodermadienediol B 71
Trichodermadienediols 111
Trichodermin 67
Trichodermol 66, 104, 110–112
Trichodiene 91, 94, 95, 97–101, 112, 113
Trichodiene synthase 99, 101
Trichodiol 92–94, 101, 102, 113
9β-Trichodiol 93
9-epi-Trichodiol 92, 94
Trichodiol A 93
Trichothecane 64
Trichothec-9,12-dienes 65, 77
Trichothecene relatives 90–93, 95, 97, 102
Trichothecenes 63–65, 73, 76, 90, 96–98, 101–103, 109, 112
Trichothecin 76, 90, 104
Trichothecinol A 76, 104
Trichothecinol B 68
Trichothecinol C 68, 104
Trichothecium roseum 66, 68, 73, 75, 76, 90–93, 96–99, 101, 102, 104
Trichothecium sp. 96, 102
Trichothecium viride 96
Trichothecodiol 66, 104
Trichothecolone 75, 98–101, 104, 105
Trichotriol 92, 94, 102, 104
9β-Trichotriol 93
9-epi-Trichotriol 92, 94
Trichoverrin A 72
Trichoverrin B 72, 111
Trichoverrin C 73
[iso]Trichoverrin C 72
Trichoverrins 111, 112
Trichoverritone 73
Trichoverroids 64, 97, 109, 111
Trichoverrol A 71
Trichoverrol B 71
Trichoverrols 111
Trifluoromethanesulfonic acid 31
3,5,6-Trihydroxyindole 163
Trimethylaluminum 41
Trimethylpsoralen 166
Trimethylsilylmethyltriflate 44
Trimethylsilyloxybutadiene 46
T-2 Triol 70
Tryptophan 161–163, 169–171
L-Tryptophan 162

Tryptophan pyrrolase 168–171
Tryptophan pyrrolase activity 164
d-Tubocurarine 53
Tyrosinase 133, 136, 139, 151–155, 158, 159, 160, 162, 163, 167–171
L-Tyrosinase 169
Tyrosinase activity 154–156, 167, 169
Tyrosinase oxidation 152
Tyrosine 133–136, 139, 151–153, 158, 159, 161–163, 165, 169–171
L-Tyrosine 162
Tyrosine aminotransferase 168–171
Tyrosine hydroxylase 169
Tyrosine-melanin 137

Udenfriend reaction 161, 162
Udenfriend system 161, 162
Urinary anthranilic acid 168
UV light 147, 148, 157

Verrol 70, 111
Verrucarin A 65, 79, 84, 96, 97, 111
Verrucarin A glucoside 79, 89
Verrucarin B 79, 84
Verrucarin H 87
Verrucarin J 65, 79, 84, 97
Verrucarin K 79, 84, 90, 112
Verrucarin L 79, 84, 112
Verrucarins 65, 84, 90, 112
Verrucarol 64, 66, 98, 111
Verticinimonosporium diffractum 85
Verticinimonosporium sp. 97
Vertisporin 81, 85, 90
Vitamin B 154
Vitamin D_3 155
Vitiligo 133, 134, 146, 157, 164–166, 168–171
Vomitoxin 75, 93, 98, 104, 106

Wadsworth-Emmons method 49
Wadsworth-Emmons reaction 33
White leprosy 164
Wilsonine 18, 53, 54

Xanthotoxin 101, 166
Xenopus leivis 167
Xylene 26

YM-47524 81, 87
YM-47525 81, 87

SpringerJournals

Amino Acids

The Forum for Amino Acid and Protein Research

Editors-in-Chief

G. Lubec, Vienna
F. H. Leibach, Augusta, GA
M. Herrera-Marschitz, Stockholm

and an **International Editorial Board**

Amino Acids publishes contributions from all fields of amino acid and protein research: analysis, separation, synthesis, biosynthesis, cross linking amino acids, racemization/enantiomers, modification of amino acids as phosphorylation, methylation, acetylation, glycosylation and nonenzymatic glycosylation, new roles for amino acids in physiology and pathophysiology, biology, amino acid analogues and derivatives, polyamines, radiated amino acids, peptides, stable isotopes and isotopes of amino acids.

Applications in medicine, food chemistry, nutrition, gastroenterology, nephrology, neurochemistry, pharmacology, excitatory amino acids are just some topics to be listed. We also encourage the submission of papers of interdisciplinary borderlines.

Subscription Information
2007. Vols. 32+33 (4 issues each). Title No. 726
ISSN 0939-4451 (print), ISSN 1438-2199 (electronic)
EUR 1148,– plus carriage charges

View table of contents and abstracts online at: **springer.at/amino_acids**

P.O.Box 89, Sachsenplatz 4–6, 1201 Vienna, Austria, Fax +43.1.330 24 26, books@springer.at, **springer.at**
Haberstraße 7, 69126 Heidelberg, Germany, Fax +49.6221.345-4229, SDC-bookorder@springer.com, springer.com
P.O. Box 2485, Secaucus, NJ 07096-2485, USA, Fax +1.201.348-4505, service@springer-ny.com, springer.com
Prices are subject to change without notice. All errors and omissions excepted.

SpringerChemistry

Fortschritte der Chemie organischer
Naturstoffe / Progress in the Chemistry
of Organic Natural Products

Edited by W. Herz, H. Falk, G. W. Kirby

Volume 84

F.-P. Montforts, M. Glasenapp-Breiling
Naturally Occurring Cyclic Tetrapyrroles

D. G. I. Kingston, P. G. Jagtap, H. Yuan, L. Samala
The Chemistry of Taxol and Related Taxoids

2002. VIII, 253 pages. 12 figures.
Hardcover EUR 164,95
Reduced price for subscribers to the series: EUR 148,–
ISBN-10 3-211-83707-8, ISBN-13 978-3-211-83707-8

Volume 85

K. Krohn
**Natural Products Derived from Naphthalenoid
Precursors by Oxidative Dimerization**

P. Messner, C. Schäffer
Prokaryotic Glycoproteins

D. P. Chakraborty, S. Roy
Carbazole Alkaloids IV

2003. X, 257 pages. 12 figures.
Hardcover EUR 184,95
Reduced price for subscribers to the series: EUR 166,–
ISBN-10 3-211-83783-3, ISBN-13 978- 3-211-83783-2

Net-prices subject to local VAT.
Recommended retail prices.

P.O.Box 89, Sachsenplatz 4–6, 1201 Vienna, Austria, Fax +43.1.330 24 26, books@springer.com, **springer.at**
Haberstraße 7, 69126 Heidelberg, Germany, Fax +49.6221.345-4229, SDC-bookorder@springer.com, springer.com
P.O. Box 2485, Secaucus, NJ 07096-2485, USA, Fax +1.201.348-4505, service@springer-ny.com, springer.com
Prices are subject to change without notice. All errors and omissions excepted.

SpringerChemistry

Fortschritte der Chemie organischer Naturstoffe / Progress in the Chemistry of Organic Natural Products

Edited by W. Herz, H. Falk, G. W. Kirby

Volume 86

Albert Gossauer

Monopyrrolic Natural Compounds Including Tetramic Acid Derivatives

2003. VII, 222 pages. 1 figure.
Hardcover EUR 130,–
Reduced price for subscribers to the series: EUR 117,–
ISBN-10 3-211-83889-9, **ISBN-13** 978-3-211-83889-1

Volume 87

T. Flessner, R. Jautelat, U. Scholz, E. Winterfeldt

Cephalostatin Analogues – Synthesis and Biological Activity

H. Budzikiewicz

Siderophores of the Pseudomonadaceae
sensu stricto **(Fluorescent and Non-Fluorescent Pseudomonas spp.)**

2004. VIII, 262 pages.
63 figures, partly in colour.
Hardcover EUR 150,–
Reduced price for subscribers to the series: EUR 135,–
ISBN-10 978- 3-211-02780-7, **ISBN-10** 978- 3-211-02780-6

Recommended retail prices. Net-prices subject to local VAT.

P.O.Box 89, Sachsenplatz 4–6, 1201 Vienna, Austria, Fax +43.1.330 24 26, books@springer.at, **springer.at**
Haberstraße 7, 69126 Heidelberg, Germany, Fax +49.6221.345-4229, SDC-bookorder@springer.com, springer.com
P.O. Box 2485, Secaucus, NJ 07096-2485, USA, Fax +1.201.348-4505, service@springer-ny.com, springer.com
Prices are subject to change without notice. All errors and omissions excepted.

Springer and the Environment

WE AT SPRINGER FIRMLY BELIEVE THAT AN INTERnational science publisher has a special obligation to the environment, and our corporate policies consistently reflect this conviction.

WE ALSO EXPECT OUR BUSINESS PARTNERS – PRINTERS, paper mills, packaging manufacturers, etc. – to commit themselves to using environmentally friendly materials and production processes.

THE PAPER IN THIS BOOK IS MADE FROM NO-CHLORINE pulp and is acid free, in conformance with international standards for paper permanency.